AQA Design and Technology

Resistant Materials Technology

GCSE

Ian Fawcett

Roger Smith

Whittle

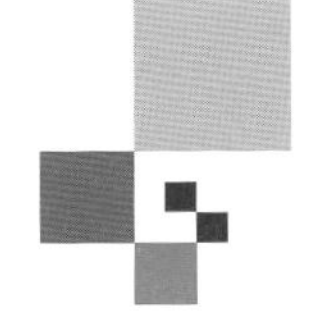

Nelson Thornes

Published in 2009 by:
Nelson Thornes Ltd
Delta Place
27 Bath Road
CHELTENHAM
GL53 7TH
United Kingdom

12 13 / 10 9 8 7 6 5 4 3

A catalogue record for this book is available from the British Library

ISBN 978 1 4085 0273 0

Cover photograph by Nelson Thornes
Illustrations by Nick Hardcastle, David Russell and Fakenham Photosetting
Page make-up by Fakenham Photosetting

Printed in China by 1010 Printing International Ltd

With thanks to Indexing Specialists (UK) Ltd.

Acknowledgements
The authors and publisher would like to thank the following companies and individuals who assisted with, or supplied photographs which appear in this book:

aliquidationcentre.co.uk; Anaheim Ice Rink; Ann Ronan Picture Library; Barratt developments; barwil.co.uk; British Olympics Committee; British Standards Institute; Canon; Corbis; domodesigner.com; Dyson; Fotalia; Getty Images; Häg Balans; henleycraft.co.uk; Chris Cullinan and Jamie Palmer / Homebase; Hyosung; Ikea; i-stockphoto; Kunsthøgskolen i Bergen; Leaf Cycles; Maclaren; Maseratti; Metropolitan Police; Microsoft Corporation; Motrax; ndv.ie; NetShelter; No Sweat; noshoes.com.au; Osann; Phillipe Stark; Radax Industries, Inc; Rigid; Robin Corps; Science Photo Library; Scott Contessa; Shutterstock; Sony; Starbucks Coffee; The Flight collection; Tommy Tippee; Topfoto; Toyota; Worldpanel; www.haggul.com; www.robbarton.com; Yamaha.

Special appreciation is offered to Clive Pryce; Jessica King; Sophie Brooks; Anita Smith; and the pupils and teachers of Banbury School, Oxfordshire; Balcarras School, Cheltenham; Great Barr School, Birmingham; Allan Rankin and the pupils of Blessed George Napier School, Banbury; Canon Slade School, Bolton; St Peters High School, Burnham on Crouch. Photo Research and commissioned photography by Ann Asquith; Dora Swick; Jason Newman; Shane Lapper; Tara Roberts of Image Asset Management & unique dimension.com; Ben Phillips Photography and Nathan Allan Photography.

Contents

Nelson Thornes and AQA

Nelson Thornes has worked in partnership with AQA to ensure this book and the accompanying online resources offer you the best support for your GCSE course.

All resources have been approved by senior AQA examiners so you can feel assured that they closely match the specification for this subject and provide you with everything you need to prepare successfully for your exams.

These print and online resources together **unlock blended learning**; this means that the links between the activities in the book and the activities online blend together to maximise your understanding of a topic and help you achieve your potential.

These online resources are available on kerboodle! which can be accessed via the internet at **www.kerboodle.com/live**, anytime, anywhere. If your school or college subscribes to kerboodle! you will be provided with your own personal login details. Once logged in, access your course and locate the required activity.

For more information and help on how to use  visit **www.kerboodle.com**.

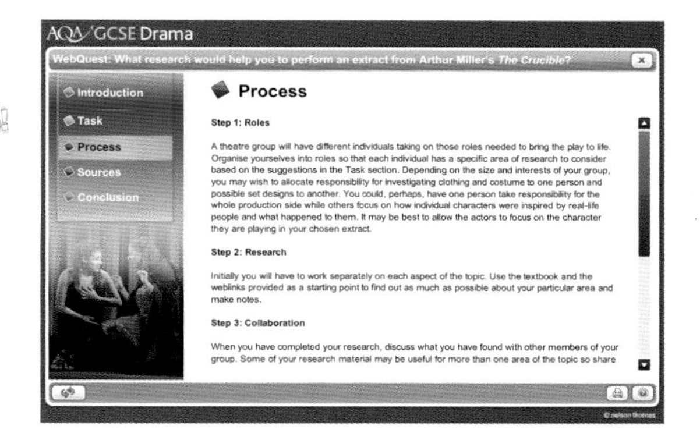

How to use this book

Objectives

Look for the list of **Learning Objectives** based on the requirements of this course so you can ensure you are covering everything you need to know for the exam.

AQA Examiner's tip

Don't forget to read the **AQA Examiner's Tips** throughout the book as well as practice answering **Examination-style Questions**.

Visit **www.nelsonthornes.com/aqagcse** for more information.

AQA examination-style questions are reproduced by permission of the Assessment and Qualifications Alliance.

The book structure

This book is divided into two units. The units correspond to the units in the AQA GCSE Resistant Materials Technology specification.

Unit 1 comprises:

- Materials and components
- Design and market influences
- Processes and manufacture.

Unit 2 is the controlled assessment unit.

By working through this book you will fully prepare yourself for the AQA Resistant Materials course. You will have all the knowledge that you will need to succeed in the external examination and you will be able to test yourself with exam-type questions.

You will be carefully led through the demands of the controlled assessment. There are examples of high quality students' work together with a detailed commentary from the moderating team.

Resistant materials

Our world is full of products that are made from resistant materials. From the moment you wake in a morning to the time you return to your bed, your whole day will be influenced and affected by products that have been designed and made from resistant materials. Indeed, the bed you sleep in, the spoon with which you eat your cereal, the toothbrush you clean your teeth with, the bus that brought you to school, the pen in your hand, the chair that you are sitting on, and many, many more products are designed and made from resistant materials.

Designing

To be a good designer it is important to understand how products have developed over time. So, in this book you will look back and learn about the main design periods in recent history. You will also look at the work of famous designers.

Most designing involves making small improvements to existing products but occasionally new, radical designs are produced. You will learn techniques that will help you become a creative designer and learn about different methods of presenting your ideas.

A designer must consider the effect their design is likely to have on others. In this book you will learn about the social, moral, environmental and sustainable issues concerned with the designing and making of products.

If you are to develop your design into a working product you need to know about materials. You will learn about the advantages and disadvantages of using a variety of materials.

Making

If you wish to make your design then you need to know about different methods of manufacture. In this book you will learn how to cut, shape, form, cast, join, clean and finish a wide range of materials.

You will learn how to work safely and how to use industrial methods of manufacture to improve the accuracy and consistency of your work.

... and finally

Resistant Materials is an exciting and very rewarding subject. Take time to learn as much as you can about the subject, particularly in Year 10. Enjoy your controlled assesment and make sure that you show your best work.

The controlled assessment tasks in this book are designed to help you prepare for the tasks your teacher will give you. The tasks in this book are not designed to test you formally and you cannot use them as your own controlled assessment tasks for AQA. Your teacher will not be able to give you as much help with your tasks for AQA as we have given with the tasks in this book.

UNIT ONE Written paper

Materials and components

In these chapters you will learn about:

- material properties
- woods
- manufactured boards
- metals
- plastics
- composite materials
- smart materials
- nanomaterials
- the sustainability of materials
- knock-down (KD) fittings and fixings
- mechanical methods of joining materials
- adhesives
- surface preparation
- applied finishes.

It is very important that designers and manufacturers have a good understanding of the different types of materials that can be used to manufacture products. Only then can they carefully select the correct material for the job. An incorrect choice of material could lead to the product breaking or malfunctioning in some way. This would be costly in terms of time, money and resources.

It is important that you can identify a range of materials and that you are able to explain clearly why they have been chosen for a particular purpose.

These chapters will explain the terminology used to describe how a material performs.

You will learn about a whole range of different traditional materials.

You will take a look at new and exciting materials.

You will look into the future of material development.

It is very important that designers, manufacturers and consumers know the source of a material and what stages it went through before it was turned into a usable material.

These chapters will explain where materials come from and how they are processed from their raw material to the final stock material, which is used to manufacture a product.

Manufacturers use a variety of methods of joining different materials together. You will learn about a wide variety of fixings, fittings and adhesives for use with a selection of different materials.

One of the most important stages in the manufacture of a product is the cleaning up or surface preparation stage. It is a stage that is quite often neglected by students and, as a result, their products look poorly presented.

These chapters will show you clearly how to prepare a surface ready for the application of a finish, for a variety of different materials.

The application of a finish is one of the last processes in the manufacture of a product and has a major influence on how a product looks. In these chapters you will learn about a wide range of finishes for a variety of different materials.

At the end of chapter 2 you will have a chance to test your knowledge using a selection of exam-style questions.

And finally, don't forget to go online where you will learn more about the topics covered in these chapters. You will also have the opportunity to test your knowledge with the exciting interactive tests!

1 Materials – metals, wood, plastics, composites and smart materials

1.1 Material properties

Different materials have different properties. These properties have to be considered by designers when choosing the most suitable and effective material to manufacture their designs. For example, properties that allow materials to be permanently shaped and formed easily can be desirable for manufacturing products in large quantities, such as plastic shampoo bottles. In contrast, materials with properties that offer the ability to withstand high forces are desirable for building bridges.

Objectives

Have an understanding of the main terms used to describe the properties of materials.

Mechanical properties

The **mechanical properties** of materials are linked to how they react to the application of a force. Some may deform in a temporary way, while for others it is more permanent.

The strength of a material is its ability to withstand an applied force without breaking or permanently bending. Materials can have different types of strength dependent on how they resist the forces being applied.

Key terms

Mechanical properties: properties of materials including strength, hardness, density, durability, toughness/brittleness, malleability, ductility and elasticity.

A *Types of strength*

Bending	Compression	Shear	Tension	Torsion
The ability to withstand forces that are attempting to bend	The resistance to forces that are trying to crush or shorten	The resistance to forces sliding in opposite directions	The resistance to forces pulling in opposite directions	The ability to withstand twisting forces

Many cutting tools such as files, saws and drills require hardness. This is the ability of a material to resist abrasive wear or scratching. Over a period of time materials can begin to weather and suffer wear and tear. Density is defined as the mass per unit volume and often has an effect on other properties, such as the hardness. Durability is the ability to resist this wear and tear. This is important because some actions can lead to the weakening of the material. These actions include corrosion (rusting) or ultraviolet light from the sun, which can cause brittleness in some plastics and can cause wood to deteriorate.

Toughness is the ability of materials to withstand sudden shocks or blows without breaking. This is something required in tools, such as hammers. Brittleness is the opposite of toughness where materials have little or no resistance to the application of a force and break very easily. Acrylic is a particularly brittle plastic unless heated.

Other important properties when shaping and deforming products are malleability, elasticity and ductility. Malleability is the ability of materials to be permanently deformed in all directions by the action of hammering, rolling or pressing. Elasticity is the ability to flex and bend when subjected to a force but regain normal shape when forces are removed. Ductility is a material's ability to be cold deformed by being pulled or 'drawn' into thinner sections or wires without breaking (Diagram **B**). The application of heat to a number of materials can alter their malleability and ductility dramatically, as seen in many types of plastic.

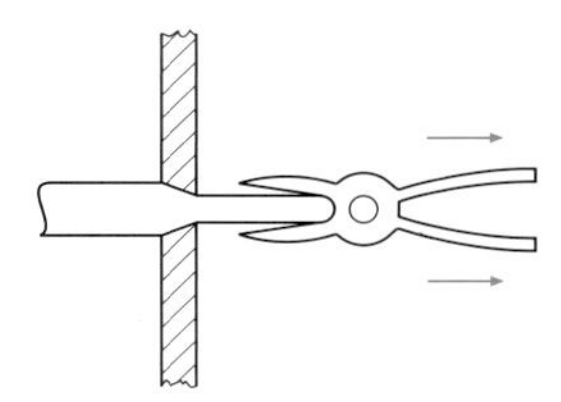

B *Ductility: a drawplate being used to pull wire through a graded hole*

Physical properties

Many of the **physical properties** of materials are unaltered by the application of forces or by the level of heat, as seen in some plastics.

Fusibility is the ability of materials to change into a liquid material at a certain temperature. This is a very important feature where materials need to be melted to carry out **fabrication** processes, such as welding and soldering, as well as forming processes, such as casting and moulding.

Thermal conductivity is the ability to allow heat to pass through a material. Electrical conductivity is the ability to allow electricity to pass through a material. Metals tend to have very good thermal conductivity while non-metallic materials have poor conductivity and are referred to as insulators. Thermal insulators can be used for preventing both heat loss and gain, for example, a polystyrene drinking cup. Good electrical conductivity is seen in most metals, particularly gold, silver and copper. Excellent electrical insulators are plastics, ceramics and wood. Some materials such as silicon are classed as semi-conductors, which means that they are normally poor conductors but under certain conditions allow a current to flow through them.

Increasingly, environmentally friendly materials are selected. They can usually either be reused or recycled and are biodegradable so that they have no negative impact on the environment.

Remember

All ductile materials are malleable but not all malleable materials are necessarily ductile.

Key terms

Physical properties: properties of materials including fusibility, conductivity and environmental friendliness.

Fabrication: the joining together of pieces, whether or not they are the same material.

links

See page 26, Sustainability of materials and page 70, Sustainability and environmental issues, for further information about sustainability.

Activities

1. Devise some simple tests that you can carry out in the school workshop to determine the different properties of a range of materials.
2. Use a graph to present the results from your tests.

AQA Examiner's tip

- You need to understand the important terms (in bold) used to describe properties of materials.
- You should also be able to describe each term in relation to a variety of materials through the use of detailed notes and sketches.

Summary

Mechanical properties of materials are linked to their reaction to force.

Many of the mechanical properties of materials often influence their selection for use in manufacturing products.

Materials also have certain physical properties that can be matched to the function of products.

1.2 Wood

Objectives

Know about the different types of wood.

Be aware of the characteristics of different types of wood and be able to visually identify a number of wood types.

Wood is probably the oldest natural material to be used by humans as both a decorative material and a structural material. This is due to its versatile properties. Today, wood is used for a wide range of activities, but in recent years we have begun to recognise that there is not an endless supply of timber. The challenge is to preserve resources, especially the slower growing trees. Due to improved forestry management, timber production is considered sustainable, which means that any tree that is cut down will be replaced by planting another tree.

The slowest growing trees (approximately 100 years) tend to be from tropical rainforests while the quickest growing trees (30 years) are increasingly grown in managed forests. Trees not only provide a valuable supply of timber, they also importantly convert carbon dioxide to oxygen, provide habitats for wildlife and contribute to the natural beauty of the countryside.

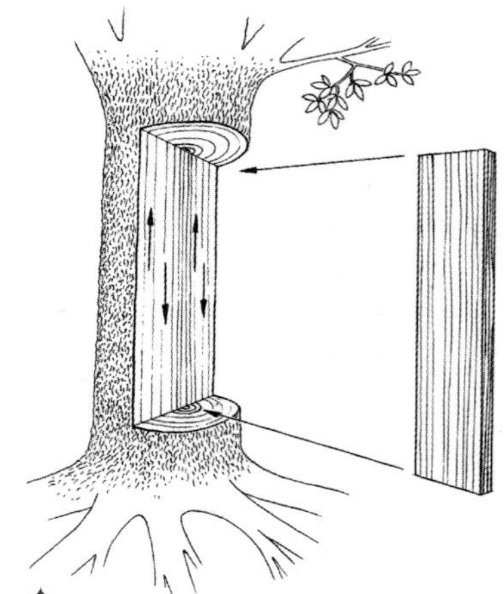

A *Structure of a tree*

Types of wood

There are two basic types of wood: **hardwood** and **softwood**.

Hardwoods

Hardwoods grown in the UK tend to be from broad-leafed, deciduous trees that lose their leaves each autumn (Photo **B**). Beech, oak and ash are examples of hardwood trees grown in the UK. Hardwoods grown in the rainforests include teak and mahogany.

Softwoods

Softwoods come from conifers, which are evergreen trees. Most conifers keep their needles throughout the year (Photo **C**). Large amounts of softwood such as pine and cedar are imported into the UK from Scandinavian countries, while the UK produces about 10 per cent of its own softwood in plantations and forests.

Key terms

Hardwood: timber that tends to be from slow growing, broad-leafed trees.

Softwood: timber from quick growing conifers.

B *A hardwood tree*

C *A softwood tree*

Conversion

Conversion means taking the trunk of the tree and sawing it into planks of usable size. There are two methods of conversion, through-and-through (basic) and quarter/radial sawn, a more complex method which can increase the stability of the wood and enhance its appearance.

- Through-and-through sawing (Picture **D**) is the simplest, quickest and cheapest method of conversion and can be used to produce planks of various thickness.
- Quarter or radial sawn wood (Picture **E**) is more expensive as the log must be cut into quarters and there can be some waste wood. This method produces the most attractive and strongest timber.

The patterns seen on wood are called the 'grain'. The direction of the grain in a piece of wood determines its strength.

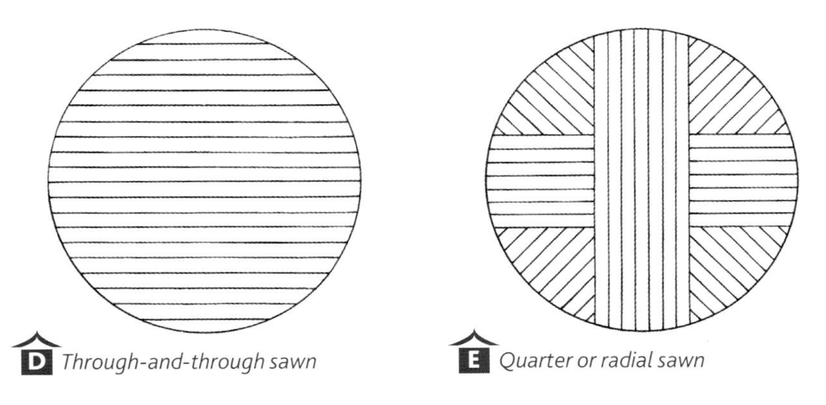

D *Through-and-through sawn*

E *Quarter or radial sawn*

Hardwoods

F *Types of hardwood*

Name	Origin	Colour	Properties and working features	Uses	Photograph
Beech	Europe	Pinkish	Close-grained, hard-wearing and strong. Finishes well	Furniture, toys, kitchen utensils	
Oak	Europe	Light brown	Tough and durable. Finishes well but stains in contact with steel	Furniture, garden furniture	
Ash	Europe	Pale cream	Tough and flexible – can be bent if steamed. Open-grained	Tool handles, laminating, sports equipment	
Mahogany	Equatorial countries	Reddish pink brown	Durable and easy to work but with some difficult interlocking grain. Alternatives woods include sapele and utile	Outdoor furniture, shop fittings, veneers	
Teak	Indian sub-continent	Golden brown	Strong and durable. Resistant to moisture. Colour darkens when exposed to light	Outdoor furniture, boats, science laboratories	

Softwoods

G *Types of softwood*

Name	Origin	Colour	Properties and working features	Uses	Photograph
Pine	Europe, Scandinavia	Cream/pale brown	Easy to work, relatively cheap and readily available. May contain knots, which weaken the wood. May also contain resin pockets, which darken with time and add to the aesthetic quality	Mainly building frames and construction Needs protection if used outdoors	
Cedar	Canada, USA	Dark reddish brown	Light in weight but not very strong, expensive, durable against moisture due to natural oils	Outdoor buildings, external wood panelling (cladding)	

Rough sawn and ready-machined wood

After conversion, wood is further converted into standard sizes and sections ready for sale. Wood is available in two forms: rough sawn and ready-machined.

- Rough sawn planks and boards are available in different thicknesses, lengths and widths.
- Ready-machined wood has been planed and is either PBS (planed both sides) or PAR (planed all round). The size of this timber is described by the original sawn size but is in fact approximately 3 mm smaller due to planing.

Some commonly available, standard sizes and shapes are as follows:

- A **plank** is over 40 mm thick and 200–375 mm wide.
- A **board** is less than 40 mm thick and 75 mm or more wide. Common thicknesses are 12 mm, 16 mm, 19 mm, 22 mm and 25 mm.
- **Squares** are sections of wood. The most common sizes are 25 mm x 25 mm, 38 mm x 38 mm and 50 mm x 50 mm.
- **Moulding and strips** are made in different sizes and shapes and are sold by length (0.9–2 m). The one used most commonly in the school workshop is dowel rod, which is available with diameters of 3 mm, 4 mm, 5 mm, 6.5 mm, 8 mm, 9 mm, 12 mm, 16 mm, 19 mm and 25 mm. Most dowel rod is manufactured from a hardwood called ramin.

Activities

1. Use samples of wood, given to you by your teacher, to devise some simple tests that will help identify some of the properties of wood.
2. Find five products made from wood that are used in your home/school (building frames, furniture, utensils, etc).
3. Sketch each product, name the wood used and a give a reason why each wood has been chosen.

Summary

Wood is a highly versatile material that has been used for centuries.

Trees are usually converted in a way to increase both strength and appearance.

Timbers can be selected for use according to their properties.

Woods are available in a variety of shapes, sections and sizes.

AQA Examiner's tip

- You should be able to name a number of individual woods and their uses.
- You should also be able to name individual woods from photographs or from looking at a range of wooden products.

1.3 Manufactured boards

Manufactured boards are made by converting logs into a variety of forms and then gluing them together to create sheet materials. The main reason for doing this is to produce large, flat sheets of timber that are stronger and more stable than conventional wide boards of softwood and hardwood. This process often uses more of the tree and therefore can be used to produce large boards of timber more economically. There are a number of additional advantages of manufactured boards, particularly how they produce a flat surface to which it is easy to add decorative **veneers**. They also provide a material that has uniform strength and good strength/weight ratio.

Producing veneers

There are two basic methods of cutting logs to produce veneers: rotary cutting (Picture **A**) and slice cutting (Picture **B**).

- Rotary cutting acts like a pencil sharpener. It produces a continuous roll of veneer that is then cut into sheets. This method is used for producing plies for plywood as the veneers are often plain-looking.
- Slice cutting produces flat sheets of veneer that often have an interesting grain pattern and are used as decorative finishing veneers.

Good uses of manufactured boards

Manufactured boards are used a lot in areas such as kitchens, flooring and furniture (Photos **C** and **D**) when a decorative or protective surface has been added.

C *BJURSTA cupboards by IKEA made with medium-density fibreboard (MDF) and a stained ash veneer*

D *FIRA storage unit by IKEA made from birch plywood*

Objectives

Know the different types of manufactured boards.

Be aware of the characteristics of different types of manufactured boards and be able to visually identify a number of manufactured board types.

Key terms

Veneer: a thin section of timber that is cut from a log and then used to produce plywood, or is glued on top of a cheaper material.

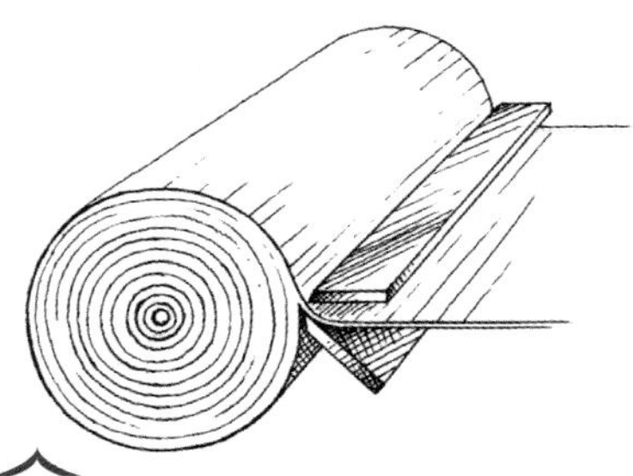

A *Rotary cutting*

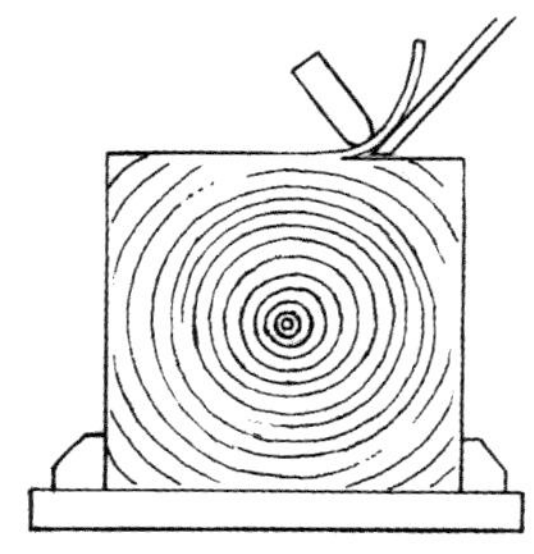

B *Slice cutting*

Activity

1. Carry out some research to help you produce a chart or table that shows the variety of methods used on the edge of manufactured boards to either protect them or make them more attractive.

Types of manufactured board

E

Name	Manufacturing process	Properties	Uses
Plywood	Veneers glued together with each layer having the grain at right angles to the previous one. There are always an odd number of layers	Flat, stable boards with constant strength and thickness. Interior and exterior varieties are available	Furniture, toys
Blockboard	Strips of softwood glued side by side and then a covering veneer added on both sides	A very strong and stable board	Furniture
Chipboard	Thousands of tiny chips of timber are mixed with glue and compressed into sheets	A cheap board that is difficult to join and is not very strong. Easily damaged by moisture. Edges are easily damaged and need some form of protection	Often in kitchens when a protective, decorative surface has been added
Medium density fibreboard (MDF)	Tiny particles of timber are glued together and then compressed with a resin adhesive to produce large, dense, solid boards	A solid and stable board that has a wide range of uses and can be produced economically. Easily damaged by moisture	Furniture, kitchen units, flooring
Hardboard	Pulped wood is treated to produce a mass of fluffy brown fibres that are then mixed with bonding agents. This is then compressed between heated sheets to produce a flat sheet, which is rough on one side and smooth on the other	A cheap flat board but with no grain and equal strength in all directions. Standard grade absorbs moisture easily but can have oil added to increase resistance to moisture	Flexible so needs supporting Cupboard backs, drawer bases and sub-floor preparations

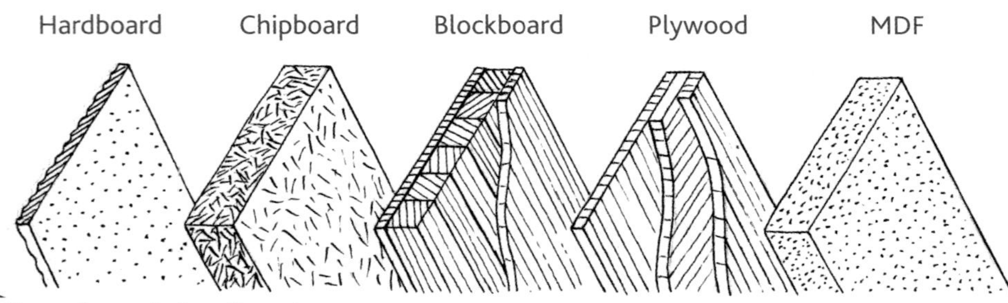

F *Types of manufactured board*

Activities

2. Manufactured boards are available in a variety of different sizes, qualities and finishes. Carry out some research to compare the cost and availability of each type of manufactured board.
3. Describe the advantages and disadvantages of using manufactured boards in your project.

Summary

Manufactured boards are essential for mass-produced furniture.

Manufactured boards that are finished with a decorative veneer make good economic use of more expensive and decorative hardwoods.

The properties of manufactured boards influence their selection and use.

AQA *Examiner's tip*

- Be able to name a number of individual manufactured boards and give examples and reasons for their uses.
- Be able to identify individual manufactured boards from photographs or from looking at a range of wooden products.
- Be able to list the advantages and disadvantages of manufactured boards when used in a variety of items.

1.4 Metals

Just like timber, metal has been used by designers for centuries and it is now often referred to as a traditional resistant material. Many of the precious metals, such as gold, have been valued and treasured for many years and were often used to produce decorative objects that were a sign of wealth. They were also used as currency but have now been replaced by cheaper alloys and paper. More recently, pure metals have increasingly been combined to produce **alloys** that have more useful properties and characteristics.

Metals are rarely found as pure metal but are normally dug from the ground as an **ore**. The metal is often extracted using high temperatures in a furnace or smelter. This process often uses high levels of energy and the outcome is a liquid metal that is processed again to produce metal forms such as sheets, tubes, rods, H-sections, etc. Some metals are extracted from ore by a chemical process.

Objectives

Know different types and names of metals.

Be aware of the characteristics of different types of metal and be able to visually identify a number of metal types.

Key terms

Alloys: metals formed by mixing together two or more metals to produce a new metal that has improved characteristics/ properties.

Ore: a solid, natural material from which metal can be extracted.

Ferrous: group of metals that contain iron and varying amounts of carbon. They are normally magnetic.

Non-ferrous: group of metals that do not contain iron.

A *Metal casting*

Ferrous and non-ferrous metals

There are two basic types of metal: **ferrous** and **non-ferrous**.

Ferrous metals contain iron and small amounts of other metals or elements, which are added to produce alloys with different properties. Almost all ferrous metals are magnetic. One major disadvantage of many ferrous metals is that they rust and therefore protective coatings need to be added, such as enamel, paint or galvanising. Stainless steel is one form of a ferrous alloy that is widely available and has anti-corrosion properties.

Non-ferrous metals contain no iron and therefore do not rust like ferrous metals and many do not require a protective coating. This makes them very useful for jobs where the metal might come into contact with moisture. Some non-ferrous metals can be expensive and most have very versatile characteristics. They are used in a wide range of products from food containers to electrical equipment.

Ferrous metals

B *Types of ferrous metals*

Name	Properties	Composition	Uses	Notes
Cast-iron	Hard skin with soft core, strong under compression but cannot be bent or forged	Iron and 3.5% carbon	Heavy crushing machinery Car brake discs Vice or machine parts	Variety of alloys available
Mild steel	Tough, ductile and malleable. Poor resistance to corrosion; cannot be hardened and tempered	Iron and 0.15–0.35% carbon	General purpose materials: nails, screws, car bodies, furniture frames, building frames and structures (Photo **D**)	Widely available at moderate costs. Generally recyclable
Medium carbon steel	Strong and hard but less ductile, tough or malleable	Iron and 0.4–0.7% carbon	Garden tools, springs	
High carbon steel	Very hard, difficult to cut and less ductile, tough or malleable	Iron and 0.8–1.5% carbon	Screwdrivers, scissors, chisels	
Stainless steel	Hard and tough but very good resistance to corrosion	Alloy of steel that also includes: chromium nickel magnesium	Sinks (Photo **E**), cutlery, work surfaces, dishes	A very tough material that is difficult to file and cut
High speed steel	Very hard, resistant to friction and can only be sharpened by grinding	Medium carbon steel alloy that also includes: tungsten chromium vanadium	Lathe cutting tools Drills Milling cutters	

Non-ferrous metals

C *Types of non-ferrous metals*

Name	Properties	Composition	Uses	Notes
Copper	Malleable, ductile and tough. Good conductor of heat and electricity. Easily joined. A green **patina** (known as 'verdigris' on copper) forms when exposed to the air or water for long periods of time	Pure metal	Plumbing fittings (Photo **F**) Electrical equipment Roofing	Can be quite expensive but polishes well. Used as a decorative roof finish due to green verdigris
Aluminium	Strong for its weight. Light, soft and ductile. Good conductor of heat and electricity. Difficult to join	Pure metal but often used as an alloy	Kitchen foils Cooking pans Window frames (usually as an alloy)	A non-toxic material that polishes well
Zinc	Very weak and heavy. Resists corrosion but very difficult to work. Often used to electro plate and hot dip ferrous products	Pure metal	Galvanising products to protect from rusting, for example, roofing, crash barriers, fencing	Greyish silver colour
Gold	A precious metal that is extremely malleable and ductile. Highly resistant to corrosion	Pure metal but normally used as an alloy that includes: silver, copper, platinum	Jewellery Gold leaf Ornaments Bullion Electrical contacts	The highest rating (24 carat) means the purest metal. Lesser ratings (18 and 9 carat) mean other metals are included

D Mild steel building frame

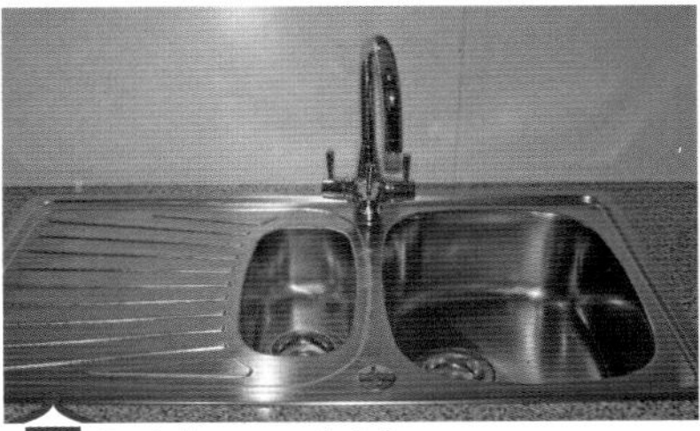
E Stainless steel sink

F Copper plumbing fixings

Alloys

Combining pure metals to produce alloys has become increasingly important in the production of more sophisticated metals required by the aeronautical, oil and space industries. Two or more metals (and sometimes other elements) are combined to produce a new type of metal that often has unique characteristics to meet very specific demands. This can be seen when a trace element, such as phosphorus (0.5 per cent), is added to bronze to improve the working/machining qualities of the metal.

Key terms

Patina: a thin coloured film that forms naturally on the surface of some metals.

G *Types of alloys*

Name	Properties	Composition	Uses	Notes
Brass	Heavy, quite hard and gold in colour Easily machined and joined by soldering	Copper 65% Zinc 35%	Castings Garden ornaments Taps and valves	Addition of zinc makes it harder than copper
Bronze	Reddish-yellow, harder than brass, corrosion resistant and casts well	Copper 80–90% Tin, aluminium or nickel in differing percentages Small amounts of phosphorus can also be added	Bearings and gears Architectural fittings Cast sculptures/statues	Addition of very small amounts of phosphorus alters working qualities and uses

Activities

1. Identify and sketch two metal objects that are made from an alloy.
2. List the benefits of using an alloy in each of your chosen objects.

Summary

Metals are available in a variety of sizes, shapes and sections.

Metals have been used for centuries for the production of objects as well as being valued and treasured.

Alloys are usually created to increase strength and improve appearance.

Different metals are selected for use due to their properties and characteristics.

AQA Examiner's tip

- You should be able to name a number of individual metals, identify them from images and describe their uses.
- You must be able to identify specific properties of metals and remember that corrosion (rusting) can affect those properties.

1.5 Plastics

Plastics mainly come from crude oil. Chemical engineers are able to mix several chemicals to produce plastic materials with almost any of the characteristics that manufacturers require. There are several types of plastic, often with complex chemical names, but many also have common names, for example, PVC, acrylic. To make them easier to identify, plastics are divided into two families: **thermoplastics** and **thermosetting plastics**.

Objectives

Know different types of plastics.

Be aware of the characteristics of different types of plastic and be able to identify a number of types of plastic.

Thermoplastics

This is the most commonly used type of plastic because it can be reshaped when reheated. The properties of this group of plastics are often changed by the addition of additives such as stabilisers, fillers, plasticisers, anti-static agents and flame retardants. Many plastics have common characteristics, such as being good electrical and thermal insulators, resisting chemical corrosion and resisting vibration and shock. Temperature changes can alter the properties of many plastics causing them to become brittle at low temperatures and lose their strength and shape at higher temperatures.

A *An oil rig*

B *Types of thermoplastic*

Common name	Working name	Characteristics	Common uses
PET	Poly ethylene terephthalate	Moderate chemical resistance, mainly used in transparent form	Bottles for drinks, food containers and in thin sheet form bonded with aluminium to form space blankets
HDPE	High density poly ethylene	Strong and stiff with excellent chemical resistance. Another popular plastic that is easily coloured with an excellent finish	Crates, bowls, buckets and pipes
PVC	Poly vinyl chloride	Good chemical and weather resistance Available in a number of forms with different properties: uPVC – tough, stiff, lightweight and can be coloured plasticised PVC – flexible	Pipes, guttering, window frames, bottles Flexible hoses, cable insulation
LDPE	Lower density poly ethylene	Tough and flexible with good chemical resistance. A popular plastic that is easily coloured and gives a very smooth finish	Detergent and shampoo bottle, toys, carrier bags and transparent packaging

Key terms

Thermoplastics: become soft and pliable when heated and can be reheated as often as required. As they cool they set again.

Thermosetting plastics: soft and pliable the first time they are heated but a chemical change takes place on cooling and they then become rigid, non-flexible and cannot be reheated and changed.

links

See page 70, Sustainability and environmental issues, for further information.

PP	Poly propylene	Lightweight, food safe with excellent chemical resistance. Good electrical insulator	Food containers, string, rope, medical equipment, kitchenware
PS	Polystyrene	Available in different forms with different properties: expanded – lightweight packaging conventional – light, hard, popular for vacuum forming in school and for food packaging toughened – increased impact strength	
HIPS	High impact polystyrene	Good stiffness and impact resistance. Lightweight	Toys, refrigerator linings
PMMA (Poly-Methyl Meth-Acrylate)	Acrylic	Tough, can be machined, but can be brittle, readily available and food safe	Light units, shop signs, car parts
Nylon	Polyamide	Hard, tough and resistant to wear. Low friction	Bearings, gears and clothing
ABS	Acrylonitrile butadiene styrene	High impact strength, lightweight, durable and scratch resistant	Kitchen products, mobile phone cases, toys and safety helmets

Thermosetting plastics

The chemical polymers that make up these types of plastics bond permanently when heated and set hard as they cool. They cannot be reheated and are usually formed into products by heating powder in shaped moulds. Many of these plastics are selected for their hardness and resistance to chemicals, oils and common solvents, but they tend to be more expensive.

C *Types of thermosetting plastic*

Common name	Characteristics	Common uses
Epoxy resins	Good chemical and wear resistance, high strength when reinforced, adhesive to many surfaces	Surface coatings, adhesives (Araldite)
Melamine formaldehyde	Rigid, good strength and hardness; scratch resistant and can be coloured	Laminates for work surfaces, tableware
Phenol formaldehyde	Very good heat resistance but very dark, hard and quite brittle	Saucepan handles and cheap electrical fittings
Polyester resins	Good heat and chemical resistance, brittle and rigid	Embedding objects, boat hulls, car parts (used with glass reinforcing)
Urea formaldehyde	Rigid, brittle, good strength, heat resistant and a good electrical insulator	Adhesives, electrical fittings such as light switches, plugs, etc.

Summary

Plastic is a highly useful material that is currently made from crude oil.

Plastics can be made to have almost any property that is required.

Activities

1. Many objects that are made from plastic were probably originally manufactured from other materials.

   Identify three such objects and then sketch each one. Explain how and why the designs differ from the original.

2. Make a list of plastic items you come into contact with during a single day.

   Produce a table that includes the type of plastic each item is made from, a sketch of the item and the reason that type of plastic was used.

AQA Examiner's tip

- You should be able to name a number of individual plastics explaining what properties they have and how this influences their use.
- You should also be able to identify individual plastics from studying a range of plastic products.

1.6 Composites

A composite material is produced when two or more materials are combined. This gives you a material with improved properties.

Objectives

Understand the characteristics and uses of different types of composite materials.

links

See Material properties on pages 10–11.

Key terms

Kit car: a self-assembly car, put together by amateur car builders.

GRP (glass-reinforced plastic)

GRP consists of strands of glass fibres that have been coated in polyester resin. It is used in the manufacture of boat hulls (Photo **A**) and '**kit cars**' (Photo **B**).

A *Sailing boat*

B *Kit car*

The polyester resin enables the GRP to be moulded into shape and bonds the glass fibres together. The addition of the glass fibre to the polyester resin improves the strength and toughness of the material. It has all the properties of a plastic:

- it can be moulded into virtually any shape
- it is waterproof; it can be given any colour pigment
- it is a good electrical and thermal insulator
- it has a good resistance to chemicals.

GRP has a rough, woven finish and therefore is usually given a 'gel coat' to provide it with a high-gloss, smooth surface.

Carbon fibre reinforced plastic

Carbon fibre reinforced plastic is a similar material to GRP. It consists of strands of carbon that have been coated in polyester resin. It is used in the manufacture of high-performance products such as racing bikes (Photo **C**) and tennis rackets (Photo **D**).

C *Track bike*

D *Tennis racket*

The carbon fibre is woven into matting and polyester resin bonds it together to form a solid shape and enables the matting to be moulded. It has all the properties of GRP but is even stronger, tougher and lighter.

Kevlar

Kevlar is a similar material to carbon fibre matting. It consists of strands of a very strong plastic material that is woven to form a mat. It is used for body armour (Photo **E**), such as the bullet-proof vests and face masks that are used by the army and the knife-proof vests that are used by the police.

The woven matting acts like a very strong net that can stop a knife attack and even a bullet! It can be combined with a plastic resin to form a solid object. It is even lighter, stronger and tougher than carbon fibre reinforced plastic.

E *Body armour*

Activities

1. Use the internet to search for as many products as you can that are manufactured using carbon fibre.
2. List the products and explain how carbon fibre enhances the product.

Summary

GRP has a smooth finish and is used to make boat hulls and kit cars.

Carbon fibre reinforced plastic is used to make racing bikes and tennis rackets.

Kevlar is a woven matting that is used to make body armour.

AQA Examiner's tip

- You will need to be able to visually identify a range of composite materials.
- You must be able to name specific composite materials.
- You must understand the specific properties of a range of composite materials.

1.7 Smart and modern materials

Smart materials are materials that have a reactive capability. This means that their physical properties change when they are influenced by something else.

Polymorph

Polymorph comes in the form of plastic granules. It has the reactive capacity to change from a solid to mouldable state when heated. When warm water (60°C) is applied to the plastic granules they melt and can then be moulded into shape. You can change the shape by reheating it, using warm water or even a hairdryer.

It is particularly useful for producing models of ergonomically designed handles.

A *Polymorph*

Thermochromic pigments

A **thermochromic** pigment can be added to a plastic before it is moulded into shape. It has the reactive capacity to change colour as its temperature changes. The plastic product will then change colour as its temperature changes when it is being used.

Designers and manufacturers have used this technology in several everyday products. Russell Hobbs make a kettle that changes colour as it boils. Tommy Tippee produce a range of baby feeding products that change colour to warn you if the baby's food is too hot.

B *Heat sensitive baby feeding spoons*

Objectives

Be able to identify a number of smart materials, understand their reactive capability and have knowledge of their uses.

Have an understanding of nanotechnology, nanomaterials, their properties and their uses.

links

See Material properties on page 10.

Key terms

Polymorph: a substance that can take different forms.

Thermochromic: having the ability to change colour as the temperature is varied.

Shape memory alloys

Shape memory alloys have the reactive capability to change their shape when heated. 'Nitinol' is a smart wire that changes length when heat is applied to it. You may have already had some in your mouth! If you have ever had a brace fitted to your teeth, the chances are that it was made from 'nitinol'. Your body heat attempted to shorten the wire, which then pulled your teeth back into shape.

'Memoflex' spectacles are made from a shape memory alloy and have the ability to return to their original shape even when they have been very badly bent.

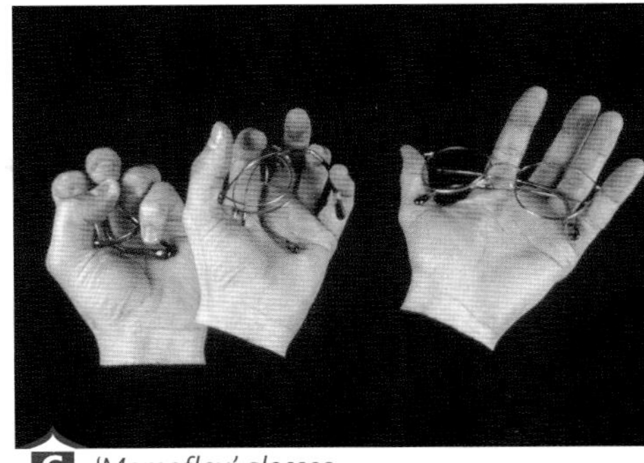

C *'Memoflex' glasses*

Nanotechnology

So far we have looked at altering a material's property by combining different materials together. Nanotechnology involves altering a material's property by changing the individual atoms that form the material.

Nanomaterials

There are many nanomaterials being developed in R&D (research and development) departments all over the world. It is seen as the future of materials development.

Nanomaterials offer great improvements in material properties. Materials are being developed that are lighter, stiffer and stronger than ever before.

links

Visit **www.nanotechproject.org** for more information about nanomaterials.

Case study

Yamaha 2009 FX Cruiser SHO

This personal watercraft uses 'NanoXcel' in the construction of its hull. This material is ultra-lightweight and is very strong. It reduces the weight of the hull, deck and liner by 25 per cent. It also gives a high gloss finish that reduces surface tension on the water and increases performance.

D *Yamaha 2009 FX Cruiser SHO*

Summary

A property of a smart material changes when it is influenced by something else.

Smart materials are chosen for different purposes according to their reactive capability.

Nanotechnology can be used to improve the quality of materials.

AQA *Examiner's tip*

- You must know what a smart material is and be able to provide at least two examples.
- You need to understand how nanotechnology improves materials.

1.8 Sustainability of materials

Plastics and metals use the earth's resources in their production. If these materials are not reused or recycled, the planet will run out of them. There is a limited amount of the **ores** (rocks or minerals) that make metals, and a limited amount of oil from which most plastics are made. If trees are not replanted as quickly as they are felled, we will run out of timber. A material is said to be **sustainable** if it can be replaced continuously, or if it can be recycled or reused indefinitely.

In the life cycle of a product, the materials need to be:

- harvested from forests (timber) or extracted from the ground (oil and ore)
- transported to a place of processing
- transported to a place of manufacture
- transported to the consumer
- transported to a place of disposal, reuse or recycling.

Each of these stages can use huge amounts of energy, causing pollution. One way to reduce the consumption of energy is to process materials and manufacture products close to the source of the material. The Diagrams **A**, **B** and **C** show the life cycle of materials, and the possibilities for making them sustainable.

Activities

1. Using the internet, research how organisations are working towards recycling glass-reinforced plastic (also known as fibreglass).
2. Use the internet to discover what plastic-based products are produced from plant matter such as wood pulp, corn (maize) and vegetables.

Objectives

Be aware of the source of a range of materials, how they are processed for use and how they can be reused, recycled or disposed of.

Understand the environmental consequences of the use of these materials.

links

See Sustainability and environmental issues on page 70, and Designers, manufacturers and product sustainability on page 72 for more on sustainability.

Key terms

Ore: a solid, natural material from which metal can be extracted.

Sustainable: something that can be replaced or reused/recycled indefinitely.

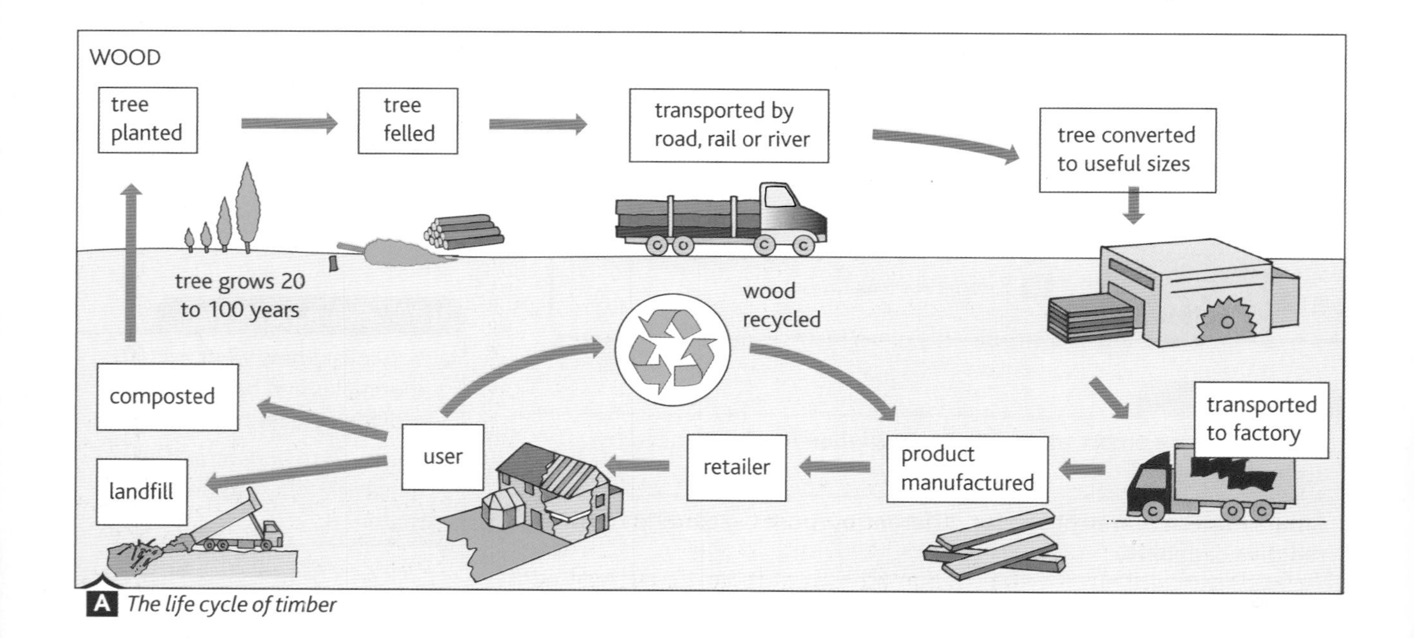

A *The life cycle of timber*

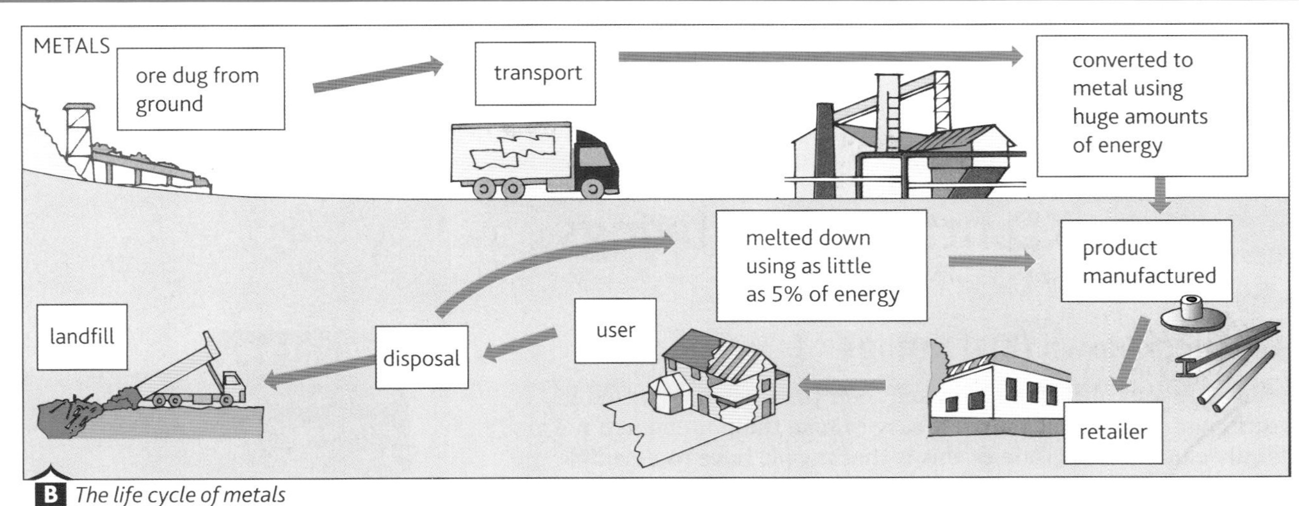

B *The life cycle of metals*

PLASTICS
crude oil extracted from ground
shipped or piped
oil refined plastics produced
product produced
retailer
user
disposal
recycled
landfill

C *The life cycle of plastics*

Sustainability of composite materials

Composite materials are produced by mixing together two or more materials. Therefore, reusing or recycling them can be complex or even impossible.

- Concrete can be crushed and used as gravel in new roads, or aggregate in new concrete.
- Car tyres can be reused by re-treading them. They can also be shredded to make floor surfaces, such as for children's playgrounds or carpet underlay, and cut up to make sandals.
- Glass-reinforced plastic (GRP) and other similar composite materials, such as Kevlar and carbon fibre, are currently almost impossible to recycle or reuse.

Sustainability of new materials

Some new materials are produced from renewable sources. For example, some plastics are now made from renewable plant-based products, which helps avoid using up the planet's supply of crude oil. These plastics will also **biodegrade** at the end of their life and can be composted.

Key terms

Biodegrade: break down naturally, decompose, rot (for example, timber).

AQA Examiner's tip

Be aware of the environmental consequences of using different materials throughout their complete life cycle.

Summary

Sustainable materials can be replaced as fast as they are used, forever.

Sustainability can be achieved by replanting forests as they are cut down, or by recycling and reusing materials.

2 Components, adhesives and applied finishes

2.1 KD fittings and fixings

Knock-down (KD) fittings

Many DIY and furniture stores sell their products flat-packed in cardboard boxes. This makes it easy to take them home in a normal family car. The downside of this is that people have to assemble the furniture themselves when they get home!

A *The perils of self-assembly furniture!*

To make it easier to assemble the products, **knock-down (KD) fittings** are used. These fittings come in a range of shapes and carry out different functions. They are designed to be used with a screwdriver or **allen key**, which is often supplied in the pack. KD fittings are used to join metal products, such as steel tube TV stands, and timber-based products such as kitchen cupboards (Photo **B**). They are also useful for joining metal to timber, and are sometimes used for plastics.

B *Kitchen cupboards are typically delivered flat-packed in boxes*

Objectives

- Have a working knowledge of a variety of types of KD fittings and fixings.
- Be able to select the correct KD fitting/fixing for a particular purpose.
- Be aware of the most common fixtures and fittings used with doors and drawers on cabinets.

Key terms

Knock-down (KD) fitting: a component that allows rapid assembly and disassembly, without damage to the parts being joined or separated. The fitting often has two parts, with one screwing or locking into the other.

Allen key: an L-shaped tool for undoing screws with a hexagon-shaped socket in the head.

C *Common knock-down fittings*

KD fitting	Photo of fitting	
Knock-down half block Screws directly to two cabinet parts to hold them at 90°. The screw holes may weaken if it is dismantled and reassembled repeatedly!		
Knock-down full block/modesty block/bloc-joint block Screws to two parts, which are then joined with a single machine screw. Good for frequent disassembly.	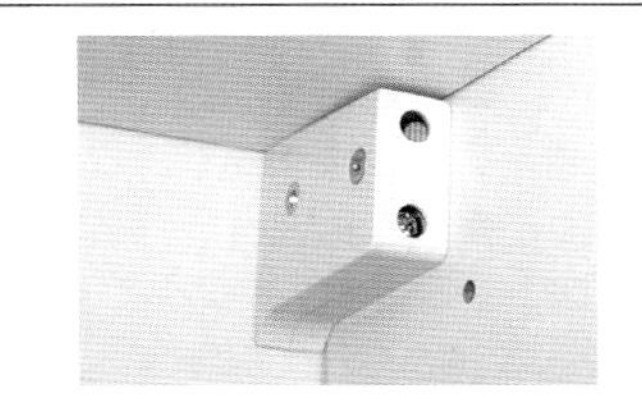	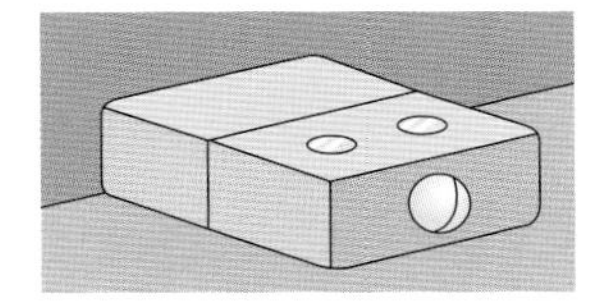 Block connector
Frame connector (also known as cross-dowel and bolt; captive barrel nut and bolt; or scan fitting) Very strong joining method because the barrel nut is fitted to a hole across one of the panels.		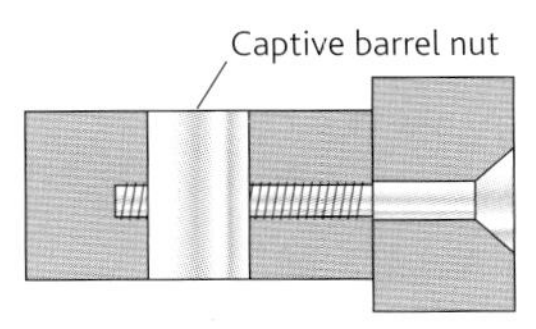 Barrel nut and bolt
Cam lock The large thread on the screw is designed to grip well in chipboard and MDF. Turning the cam pulls the parts together.	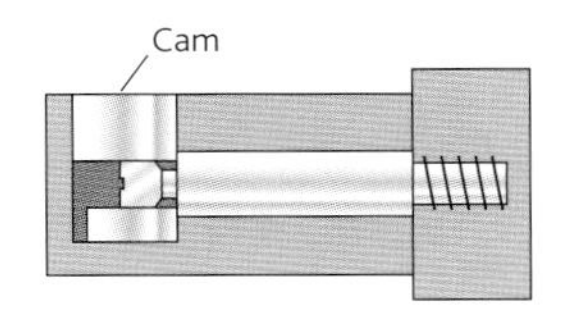 Cam fitting	
Worktop connector Fitted underneath worktops. Tightening the nuts with a spanner pulls the worktops together.	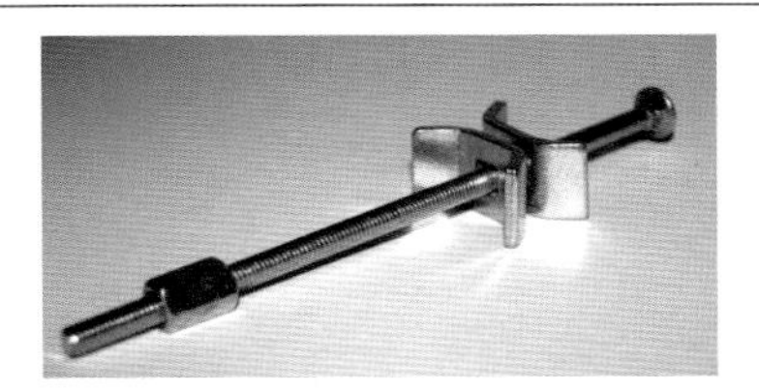	
Corner plate Allows a table leg to be fitted to a table frame without tools. A special threaded wood screw onto which a tightened wing nut holds the leg in place.		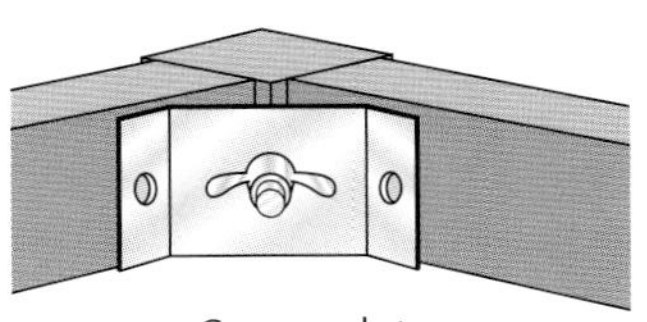 Corner plate
Captive nut/pronged tee nut The nut bites into the timber. The bolt then screws into the hard nut, rather than soft timber.	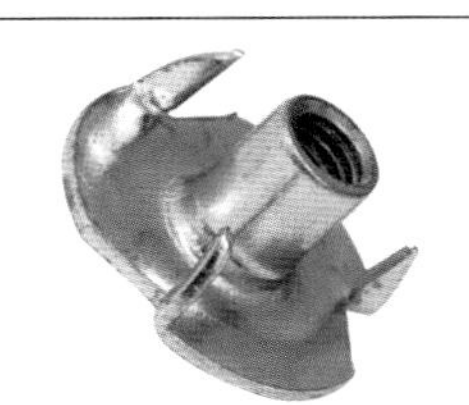	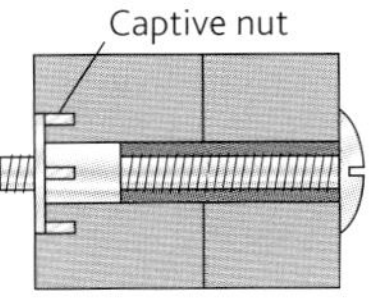 T-nut and bolt

Activity

Write a plan of installation for each knock-down fitting shown in Table **C**.

Other fixtures and fittings

There is a huge variety of fixtures and fittings. This is a small selection of the most common of these components, used for moving parts on products.

D *Fixture and fitting components*

Bales catch Fits into the edge of a door, so is almost totally hidden from view.	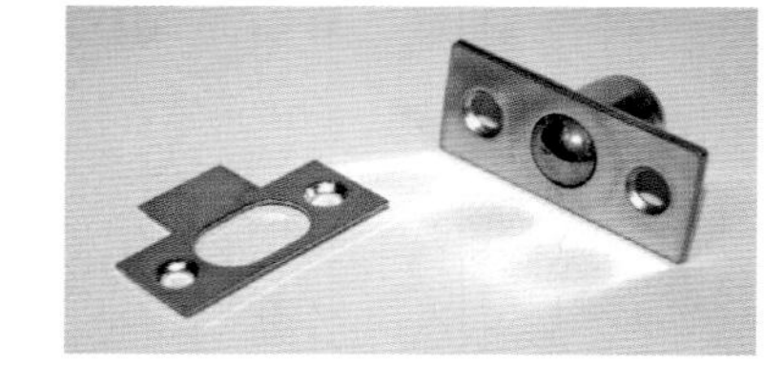
Showcase latch A decorative catch used on cabinet doors. Very popular on jewellery cases.	
Magnetic catch Simple to fit, but visible when the door is open.	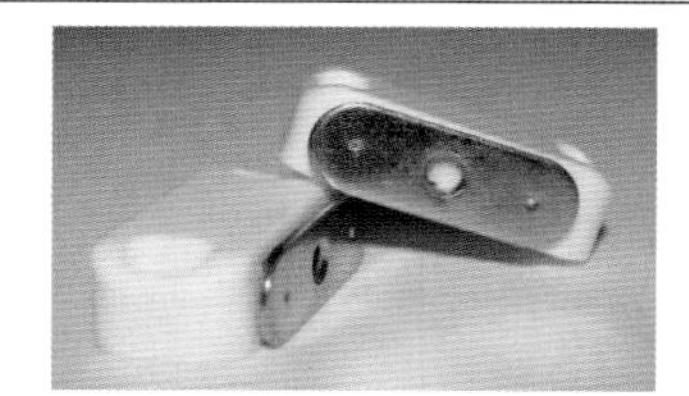
Bolt Often used to fix one of a pair of doors.	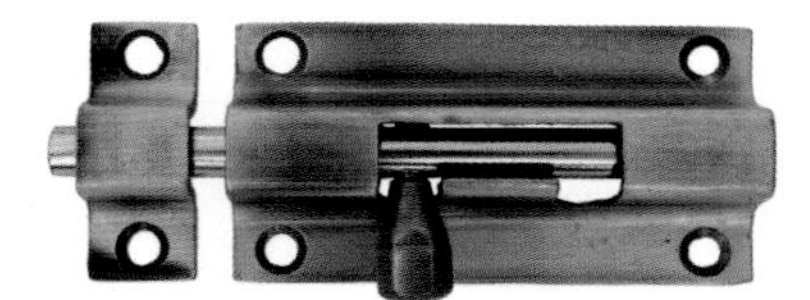
Toggle latch Used for holding lids closed. Can be used vertically or horizontally.	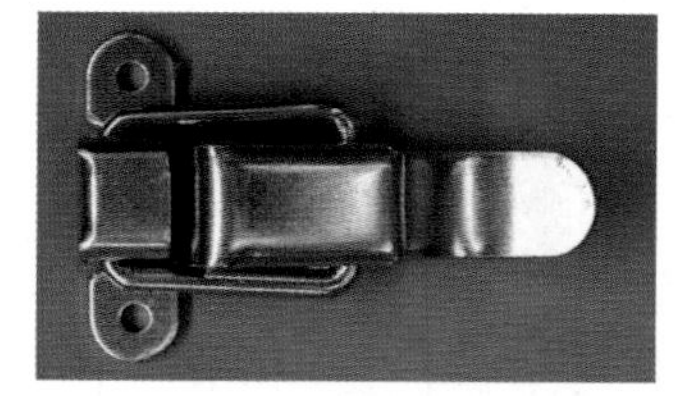
Lid stay These come in all shapes and sizes. Used for holding flaps open and for limiting door opening.	
Drawer runners Allow drawers to slide much more easily than traditional methods.	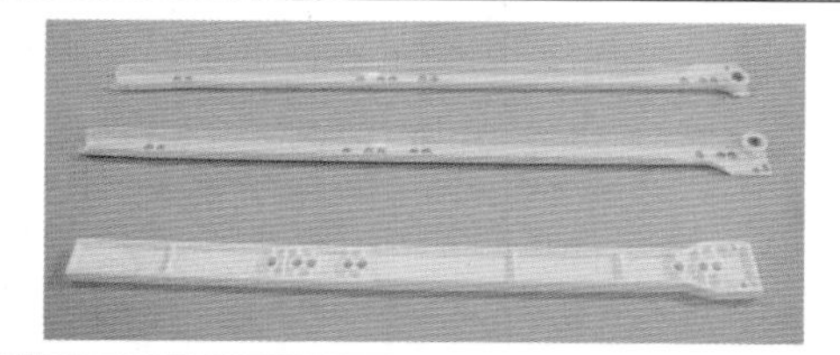

links

To see more fittings and fixtures, visit **www.screwfix.com**

Specialised fixtures and fittings can be seen at **www.ironmongerydirect.co.uk**

Hinges

Hinges come in a huge variety of shapes and sizes, and materials that are suitable for their intended environment. They allow doors or flaps to open and close. Popular, everyday hinges are shown below.

E *Commonly used hinges*

Hinge	
Concealed hinge These are used for kitchen and other cupboard doors. They are sprung so need no catch. They are easy to adjust so the door will line up well.	
Soft close device Fitted to more expensive cabinets to stop the doors from slamming shut.	
Brass butt hinge A traditional hinge that looks best if it is recessed into the frame and door.	
Steel butt hinge Cheaper alternative to the brass butt. Not weather resistant. Pressed from sheet steel.	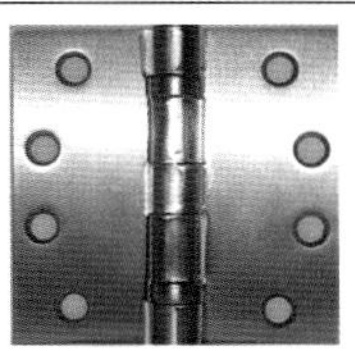
Flush hinge Simpler to fit than the butt hinges as it needs no recess.	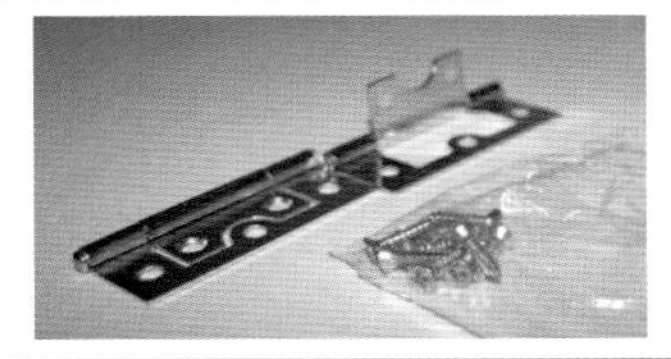
Decorative hinge Easy to fit as it is screwed to the outside of a cabinet.	
Piano hinge Long, continuous butt hinge that can be cut to the correct length.	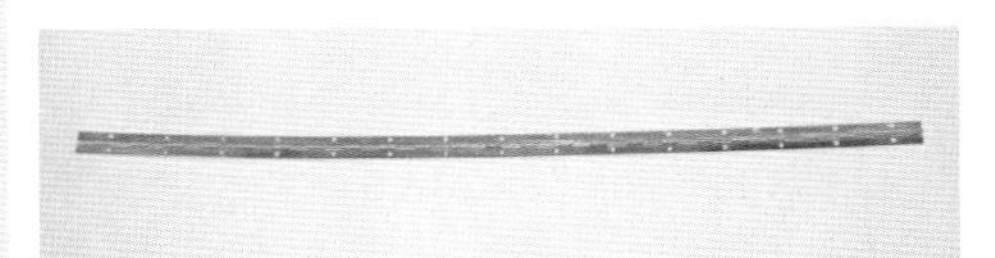
Tee hinge Useful for shed doors and gates.	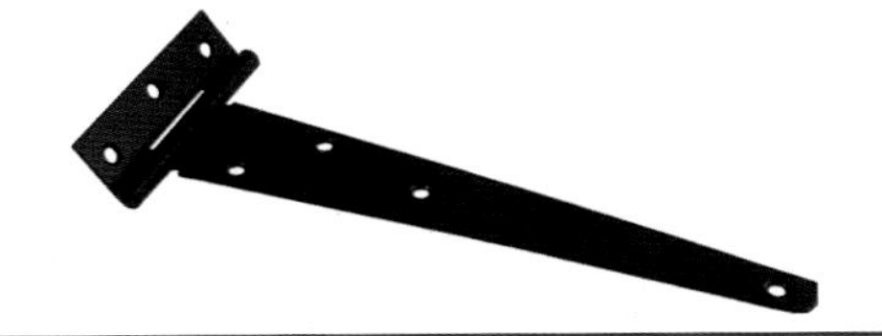

AQA Examiner's tip

- You need to learn which components are best suited to particular situations and be able to justify why you have chosen a particular component.
- Practise sketching the components in use.

Summary

Knock-down (KD) fittings allow furniture parts to be assembled easily with simple tools.

Catches, latches and bolts allow doors and flaps to be secured.

Hinges allow doors, flaps and lids to open.

2.2 Mechanical methods of joining materials

Mechanical joining systems allow parts to be dismantled at a later date without damaging the materials. This is unlike permanent methods, such as welding and gluing, where parts will be badly damaged if they are taken apart.

Mechanical (non-permanent) methods of joining materials make use of:

- nails
- screws
- nuts and bolts
- rivets.

Objectives

Know different types of screws and nails and understand how they are used.

Have knowledge of common nuts, bolts, washers and rivets.

links

See more information on permanent methods of joining on page 36, Adhesives.

Key terms

Galvanised: a coating of zinc applied to steel to stop it from rusting.

Nails

Nails are quick and easy to use. They need only a hammer, plus a nail punch if you wish to drive the head below the surface. Parts can usually be knocked apart with little damage to the surfaces. Use a block of wood held against the surface to spread the impact of the hammer when dismantling.

Nails are mostly made from mild steel, but other materials such as stainless steel, copper and aluminium are used for specific purposes. Some mild steel nails are available with a **galvanised** finish for use outdoors.

A *Types of nail*

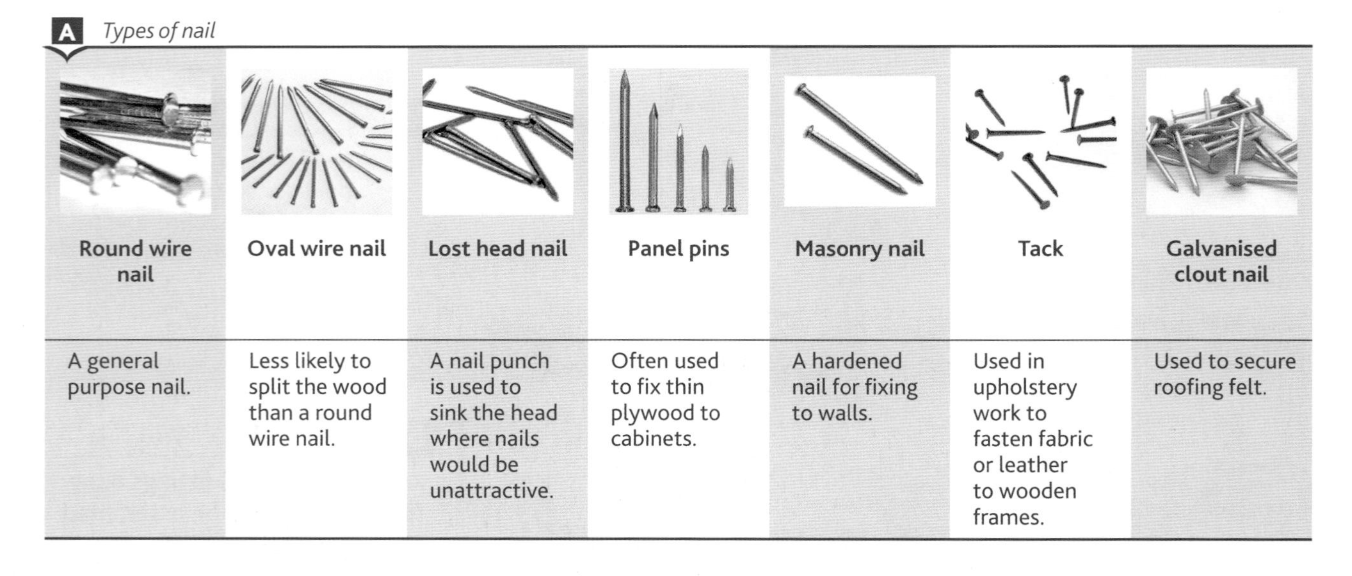

Round wire nail	Oval wire nail	Lost head nail	Panel pins	Masonry nail	Tack	Galvanised clout nail
A general purpose nail.	Less likely to split the wood than a round wire nail.	A nail punch is used to sink the head where nails would be unattractive.	Often used to fix thin plywood to cabinets.	A hardened nail for fixing to walls.	Used in upholstery work to fasten fabric or leather to wooden frames.	Used to secure roofing felt.

Screws

Wood screws can be used to join together a variety of materials. The shape of screw will vary according to what materials it is joining and where it is going to be used. It will also vary in length and diameter. A single tooth spirals around the shank of the screw. It is this that pulls the screw into the material when it is turned with a screwdriver. Cross head screws are very popular because they work well with cordless screwdrivers (Photo **B**). Traditional screws have a slotted head (Photo **C**). Screws are commonly made from steel, but brass, stainless steel and other materials are also used. Some steel screws are available zinc or chrome plated.

B *Modern cross head screws*

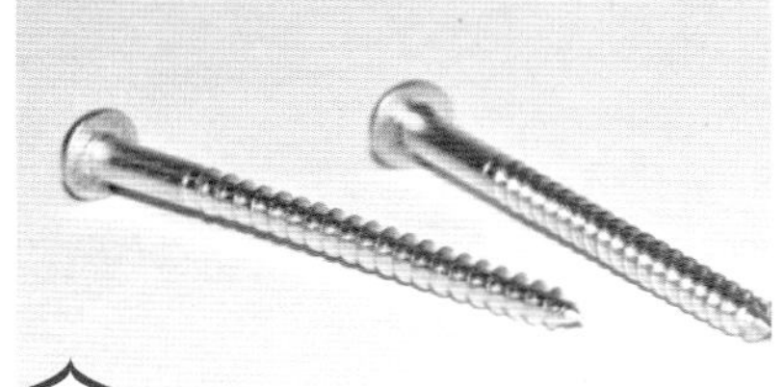

C *Traditional slotted wood screws*

Joining two pieces of wood with screws

This is the correct method for joining two pieces of wood with screws (Diagram **E**):

- In the base material, drill a pilot hole. The tooth of the screw will bite into the sides of this hole.
- In the material to be fixed to it, drill a clearance hole. The screw should be a sliding fit in this.
- Countersink the clearance hole, so that the head of the screw can be driven level with the surface of the wood.

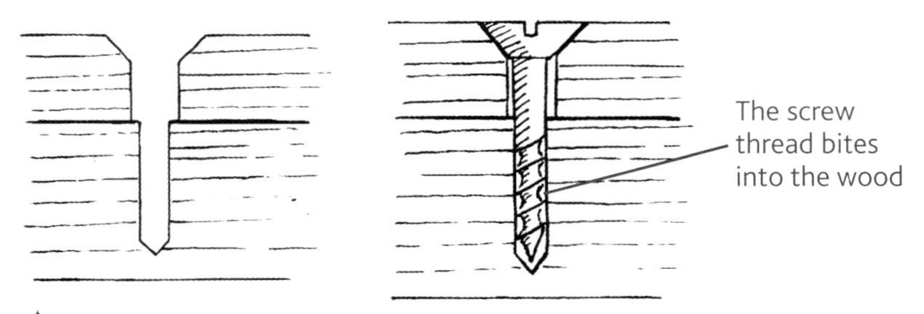

E *A countersunk screw used to join two pieces of wood*

F *A selection of specialised screws. Note the variety of head shapes and threads*

Key terms

Countersunk: a hole that is recessed, to allow the head of a countersunk screw to be driven level with the wood surface.

Remember

If you drive the nails in at different angles, it makes the joint harder to pull apart. This is called dovetail nailing (Picture **D**).

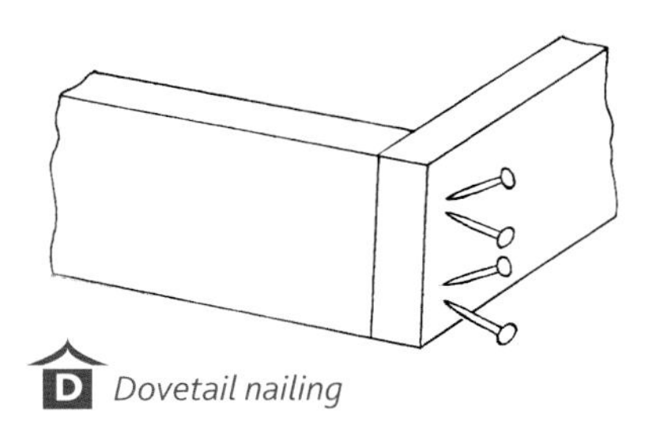

D *Dovetail nailing*

Activity

1 One of the leading suppliers of screws in the UK is www.screwfix.com. Use their website or their catalogue to:

a name the specialist screws in Photo **F**

b state what each is used for

c describe how each one is specially adapted for its function.

Nuts and bolts

The screws shown on the previous page are designed for fixing wooden parts together. For fixing metal parts together, it is more usual to use nuts and bolts. Both systems are non-permanent and they can be taken apart easily. Like screws, nuts and bolts come in all shapes and sizes, depending on their intended task.

Activity

2 Study a bicycle. List the different types of nuts and bolts used. Explain why you think that each particular component has been used.

G *Common nuts, bolts and washers*

Bolts		Nuts and washers	
Hexagon head bolt	Tightened with a spanner.	Standard hexagon nut	Tightened with a spanner.
Coach bolt	Square section under the head bites into wood to stop it from spinning.	Wing nut	Can be tightened by hand.
Pan head machine screw	Machine screws have a thread running the whole length of the bolt.	Dome nut	More attractive and safer than leaving the end of a bolt exposed.
Machine screw	Tightened with a cross head screwdriver.	Nyloc nut	Special plastic insert stops the nut from vibrating loose.
Grub screw	Tightened with an allen key to lock a part to a round shaft.	Standard washers	Washers are used under the nut to stop it sinking into the surface.

Threading

It is possible to cut a thread on a metal rod (male or external thread). This is called 'threading'.

It is also possible to cut a thread in a hole (female or internal thread). This process is called 'tapping a thread'.

Hardened steel screws are available that cut their own threads. These are called self-tapping screws and are used in computer casings and the joining of two-part moulded plastic cases of electrical components such as irons and vacuum cleaners.

Rivets

Rivets are a convenient way to join metal sheets without having to use heat (heat is needed for welding). A hole is needed for the rivet to fit into. The end of the rivet is hammered over once fitted.

Rivets cannot be undone. To remove them, they have to be drilled out.

Rivets are available with a variety of head shapes, and can be made from soft mild steel, aluminium, brass or copper.

Pop rivets

A pop rivet gun allows rivets to be fitted from one side. There is no need to hammer the end of the rivet. These are known as pop or blind rivets (Photo **H**).

H *Pop rivets*

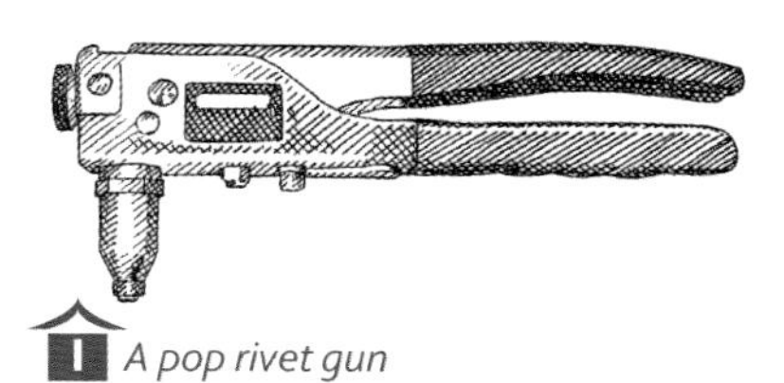

I *A pop rivet gun*

Summary

Nails, screws, rivets, nuts and bolts are all mechanical methods of joining materials together.

The materials can be separated again, usually without damage.

They are all available in a variety of shapes and sizes, depending on their intended use.

Most are made from steel, but other metals are used for specific situations.

links

Compare the methods shown on this page with permanent methods shown on pages 36–37.

AQA Examiner's tip

- You must be able to select and name appropriate fixings for different situations.
- It is also important to be able to name the tools used with different fixings.

2.3 Adhesives

Adhesives are used to bond materials together to form a permanent joint. There are a wide variety of adhesives to choose from and it is important to select the correct one and understand how to apply it safely.

Objectives

Be able to select and use the correct adhesive to bond together certain materials.

Preparation

Preparation of the material is very important. All adhesives need the material to be clean, dry and free from oil and dust if they are to achieve their maximum grip.

Some areas may need to be covered with masking tape to prevent the glue spreading over other areas.

Some adhesives require the joint to be '**keyed**'. This means that the joint should be made rough (usually done with an abrasive paper).

Key terms

Adhesives: types of glue.

Keyed: a surface is roughened to improve the strength of a joint when two surfaces are stuck together.

links

Look at the online resources to see some more unusual adhesives and techniques.

Dry run

It is always useful to have a dry run before you apply the adhesive. Put together the materials to be joined but do not bond yet with adhesive. This will allow you to check the fit of the joint and enable you to get any clamps assembled and set up ready to use.

Choosing the correct adhesive

Use the following selection grid to help you decide which adhesive to use.

Carefully follow the instructions provided with the adhesive.

You **must** make sure you observe all the safety instructions.

A *Types of adhesive*

Name	Material	Drying time	Use
PVA (Polyvinyl acetate)	Wood	4 – 24 hours	This adhesive gives a strong joint. It comes in liquid form and can be used straight from the bottle or applied with a brush. Joints need to be held together while the glue dries.
Synthetic resin 'Cascamite' 'Extramite'	Wood	6–8 hours	This adhesive is waterproof and gives a strong joint. It comes in powder form and must be mixed with water, then applied by a brush. Joints need to be held together while the glue dries.
Contact adhesive 'Evostick'	Wood, metal, plastic	Instant	This adhesive is waterproof and gives a medium strength joint. It is ideal for plastic laminates to chipboard for kitchen worktops. It comes in liquid form and is applied using a spreader. It is thinly applied to both surfaces. The surfaces are left to 'touch dry', then they are pressed together.

Epoxy resin 'Araldite'	Wood, metal, plastic	½–6 hours	This adhesive is waterproof and gives a strong joint. It comes in two tubes, a resin and a hardener. Equal amounts of resin and hardener are mixed together and applied with a spreader. Joints need to be held together while the glue dries.
Liquid solvent cement 'Tensol'	Thermoplastics	10 minutes	This adhesive is waterproof and gives a medium strength joint. It comes in liquid form and is applied with a brush. Joints need to be held together while the glue dries.
Hot glue stick	Wood, metal, plastic	On cooling	This adhesive is waterproof but is weak and only suitable for modelling or temporary fixings. It comes as a solid stick of glue. It is heated in a special gun and applied direct from the nozzle.
Cyanocrylate 'Super glue'	Wood, metal, plastic	Instant	This adhesive is waterproof and gives a medium strength joint. It comes in a liquid form and is applied direct from the nozzle.

Activities

1. Choose three types of wood adhesive from the list.
2. Perform an experiment on three test joints to find out which is the strongest.
3. Produce a report of your experiment and add pictures to illustrate what you did.

AQA *Examiner's tip*

Make sure that you can describe in detail how to use several types of adhesive. You should use both notes and sketches.

Cleaning the joint

It is important to clean off any excess adhesive as soon as possible. Excess glue will make your work look bad and may stain the surface.

Summary

There are a variety of types of adhesive for joining together different materials.

To achieve maximum grip, materials must be clean, dry and free from oil and dust before applying adhesive.

Excess adhesive should be cleaned off immediately to provide a neat finish and avoid staining.

2.4 Surface preparation

It is essential that the surface of a material is fully prepared before a finish is applied. This will vastly improve the quality of the work.

Objectives

- Understand the need for surface preparation before the application of a finish.
- Have knowledge of a range of methods of preparing different materials before a finish is applied.

links

See Applied finishes on page 40.

Preparing a wooden surface

The first stage is to get the surface and edges flat. There are several different methods for doing this. Which one you use will depend on the tools and equipment that are available to you and your skill level:

- A sharp plane (Photo **A**) will produce the flattest, smoothest surface, but does require some skill.
- Mechanical sanding using a band facer, linisher (a sanding machine), disc sander (Photo **B**) or a palm sander will produce a flat surface but will usually leave sanding marks.

Once flat, the surface should be finished by hand sanding using an abrasive paper such as glass paper and a sanding block. You should work your way through the grades of glass paper, from coarse to fine. Remember to sand with the grain to avoid scratches going across the wood and also remember to wipe away any dust created.

A *Plane*

B *Disc sander*

C *Mechanical sanding*

Preparing a metal surface

The first stage is to get the edges of the metal smooth. This is best done by **draw filing** the metal then **deburring** the edges. Draw filing makes the edges smooth (Photo **D**) and deburring ensures that there are no sharp pieces of metal on the edge.

The next stage is to clean the surface. This should be done using an abrasive paper such as emery cloth wrapped around a file. You can add oil to this process to get a better finish.

The metal should then be cleaned with white spirit to remove any grease or oil from the surface.

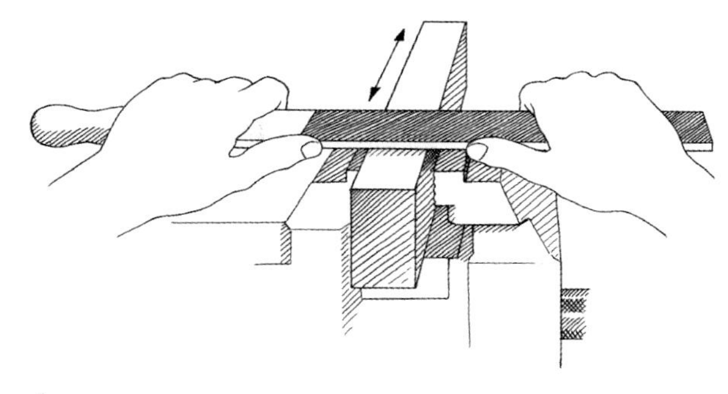

D *Draw filing*

Preparing a plastic surface

Most plastics are self-coloured and have an immaculate surface finish. You must protect this surface finish by keeping any protective coating on the plastic for as long as possible. Once the edges have been cut they will need preparing before they can be finished.

The first stage is to get the edges of the plastic smooth. This is usually done by draw filing the plastic and then using an abrasive paper such as wet and dry paper (silicon carbide paper).

Key terms

Draw filing: smoothing the edge of metal or plastic by drawing a file backwards and forwards.

Deburring: removing the sharp edge from a piece of metal by drawing a file backwards and forwards.

Abrasive papers

Abrasive papers are used for finishing materials and come in a variety of grades. The higher the grit number, the finer the grade:

- coarse glass paper and emery cloth is around 80 grit
- fine wet and dry paper (silicon carbide paper) is around 400 grit.

Activity

1 Cut out three key fobs similar to the one shown in Diagram **E**.
   a Make one out of 3 mm plywood, one out of aluminium and one out of acrylic.
   b Prepare the edges of each ready for finishing.
   c Log each stage of the process using a flowchart.

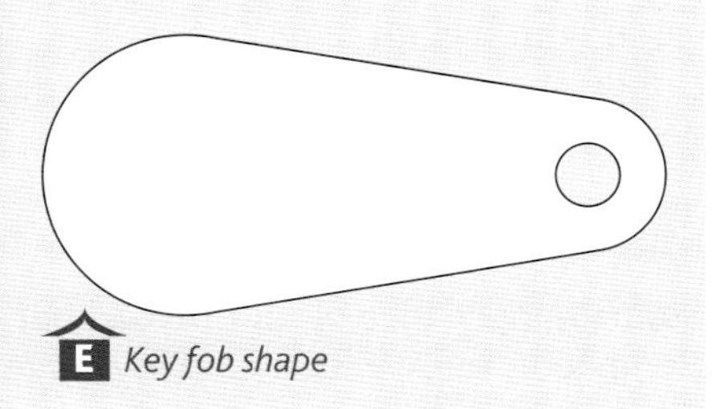
E *Key fob shape*

Summary

Surfaces must be prepared before the application of a finish.

Different methods of surface preparation are used for different materials.

AQA Examiner's tip

Make sure that you can describe in detail how to prepare surfaces ready for a finish to be applied. You should use both notes and sketches.

2.5 Applied finishes

There are two main reasons why it is essential to apply a finish to a material.

Aesthetics: applying a finish will improve the way a product looks. It can improve the natural look of the material or it can completely change its colour and texture.

Protection: applying a finish will protect a material from becoming damaged when in use.

Objectives

- Understand why a finish needs to be applied to a material.
- Have knowledge of a variety of types of finishes and how to apply them.
- Be able to select a finish and relate it to a particular material.

Finishes for wood

Wax

Wax adds a shine to the surface of wood and gives some protection. There are two main types of wax: beeswax and silicon polish.

Wax comes in a solid form and is rubbed into the surface of the wood with a cloth. Once it has dried, the wax is then **buffed** to create a natural, shiny surface. Extra coats can be applied to produce a deeper shine.

Key terms

Buffed: mechanically or hand-polished to produce a high-quality, shiny surface.

Oil

Oil adds shine to the surface of wood and provides some protection.

There a number of different types of oil finish: teak oil, Danish oil and linseed oil being the most common.

Oil comes in liquid form and is rubbed into the surface of the wood with a cloth. Extra coats can be applied to produce a deeper shine.

A *Teak patio furniture*

Stain

Stain changes the colour of wood but provides little protection. Stain is applied evenly with a cloth over the surface of the wood. Once it has dried it needs to be sealed with a sealer or varnish to create a shine and to add protection.

French polish

French polish adds a deep shine to the surface of wood and gives some protection.

This is a traditional and very skilled method of finishing wood. French polish comes in a liquid form and is a mixture of shellac (beeswax) and methylated spirits. It is applied with a 'rubber' (cotton wool wrapped in a cloth). Many layers are built up until the required shine is obtained.

Remember

Uses: wax is generally used on interior furniture, such as a dining table.

Uses: oil is generally used on interior furniture. Teak oil can be used on teak outdoor furniture (Photo **A**) as well as indoor furniture.

Uses: stain can be used to colour any wooden products but must be sealed.

Uses: French polish is generally used for high quality furniture.

Sealer

Sealer adds shine to the surface of wood and gives good protection. There are a number of different types of sealer; sanding sealer and medium-density fibreboard (MDF) sealer are the most common.

Sealer comes in a liquid form and is brushed onto the surface of the wood. Once dry, the sealer is rubbed down with fine glass paper. A wax or varnish finish can then be added.

Varnish

Varnish adds shine to the surface of the wood and gives good protection. Varnishes can also be coloured.

There are two main types of varnish: polyurethane and acrylic.

Varnish comes in a liquid form and is brushed onto the surface of the wood. Once dry, the varnish is rubbed down with wire wool. Extra coats can be applied to produce a deeper shine or a wax finish can be added.

Paint

Paint adds colour to the wood and is available in a matt, silk or gloss finish. It provides good protection.

There are three main types: oil-based paint, water-based paint and solvent-based paint.

Oil- and water-based paints are the most commonly used on wood. However, the preparation of the wood before oil-based paint can be applied is quite time consuming. Any knots in the wood need to be treated then rubbed down; a primer coat needs to be applied then rubbed down; an undercoat needs to applied then rubbed down; and only then can you apply the first top coat. Extra coats can be applied to produce a deeper shine. Water-based paints use acrylic, are non-toxic and are therefore ideal for use on children's toys.

Solvent-based paint comes in the form of a spray can. It does not leave brush strokes and is quick-drying; however, it is more expensive.

Finishes for metal

Paint

Paint adds colour to the metal and is available in a matt, silk or gloss finish. It gives good protection.

All three types of paint (oil-based, water-based and solvent-based) can be used for metal; however, solvent-based paint is the most commonly used.

A primer coat is applied then rubbed down; an undercoat is applied then rubbed down; and then the first top coat is applied. Extra coats can be applied to produce a deeper shine.

Hammerite paint is a 'one-coat' cellulose-based paint with a hammered finish. It is ideal for use on outdoor steelwork such as gates.

Lacquer

Lacquer adds a clear shine to the metal and gives good protection.

Remember

Uses: sealer is generally used on interior furniture.

Uses: varnishes can be for interior or exterior use. Polyurethane varnish is traditionally used for protecting wood on boats.

Uses: paints can be used for interior or exterior wooden products (Photo **B**).

Uses: paint can be used for interior or exterior metal products, such as radiators and garden gates.

B *Number 10, the world's most famous painted door!*

Lacquer is generally solvent-based and comes in the form of a spray can. It does not leave brush strokes and is quick drying.

Plastic dip coating

Plastic dip coating adds colour to the metal and gives excellent protection.

The metal is first heated up in an oven to 200°C. It is then dipped into a fluidising bath of powdered polythene. The polyethylene sticks to the hot metal and is left to cool.

Oil bluing

Oil bluing adds a bluey/black colour to steel and gives it some protection.

Steel is first heated to 700°C (chimney red). Then it is plunged into an oil bath. The oil sticks to the hot metal and is left to cool.

Anodising

Anodising adds vivid colour to aluminium. It gives excellent protection and hardens the surface. Anodising is an industrial process that involves the use of **electrolysis**.

C *Anodised bicycle rims*

Plating

Plating involves the coating of a metal with another metal. It changes the appearance of the metal and gives excellent protection. Plating is an industrial process that involves the use of electrolysis.

Remember

Uses: lacquer can be used for interior or exterior metal products.

Uses: plastic dip coating can be used for interior or exterior metal products.

Uses: oil bluing is most commonly used on small tools and steel components.

Uses: specialist parts for bikes are commonly anodised aluminium components (Photo **C**).

Uses: chrome plating is the most common form of plating. Bathroom taps are an excellent example (Photo **D**).

Key terms

Electrolysis: the process of coating a metal by placing it into a solution of electrolyte and passing an electric current from the donor metal to the parent metal.

D Chrome plated bathroom taps

Galvanising

Galvanising involves the coating of a metal with another metal. It changes the appearance of the metal into a mottled dull grey and gives excellent protection.

Galvanising is an industrial process that involves dipping steel into a bath of **molten** zinc.

Finishes for plastic

Plastics are usually self-coloured and self-finished. This means that a colour pigment has already been added during the manufacture of the plastic and the manufacturing process gives a very high quality surface finish.

The only process that should be required is polishing with a buffing machine or by hand with a metal polish such as Brasso.

Key terms

Molten: the state when a metal has turned to liquid by heating.

Remember

Uses: galvanising is frequently used on ships and outdoor metalwork, particularly any located near the sea.

Summary

There are two reasons for finishing materials: aesthetics and protection.

Wood can have a number of finishes applied to it: wax, stain, oil, French polish, sealer, varnish and paint.

Metal can have a number of finishes applied to it: paint, lacquer, plastic dip coating, oil bluing, anodising and plating.

AQA Examiner's tip

Make sure that you can describe, in detail, how to apply at least two types of finish for both wood and metal.

AQA

Examination-style questions

Test your understanding of materials and components by answering the following exam-style questions.

Materials and their properties

1 a) Name a suitable, specific material that has been used to make each of the products shown below. *(3 x 2 marks)*

b) In each case, give **one** reason for your choice. *(3 x 1 mark)*

AQA specimen question

A

B

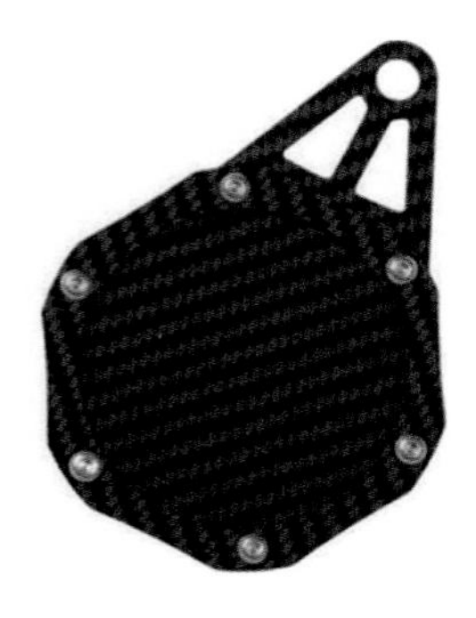

C

2 Use notes and sketches to describe how plywood gets its strength. *(8 marks)*

3 Describe what is meant by the term 'smart material'. *(2 marks)*

4 Give examples of two smart materials and their uses. *(4 marks)*

5 Describe the material GRP. *(6 marks)*

6 Explain the effects that nanotechnology has on the performance of new materials. *(4 marks)*

7 Describe how wood is obtained from trees. *(6 marks)*

8 Discuss the sustainability issues of producing and using products made from wood. *(8 marks)*

Components, adhesives and applied finishes

9 Study the fixings shown below.

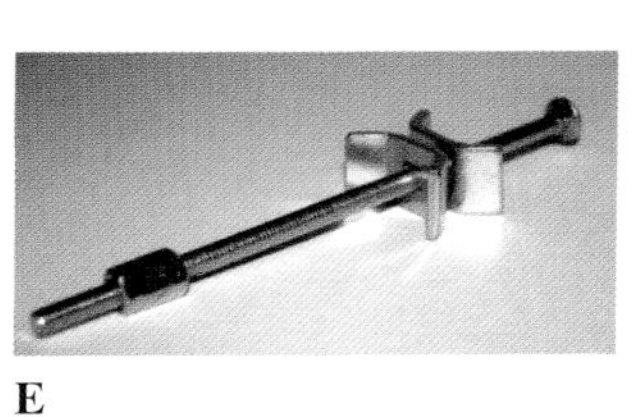
E

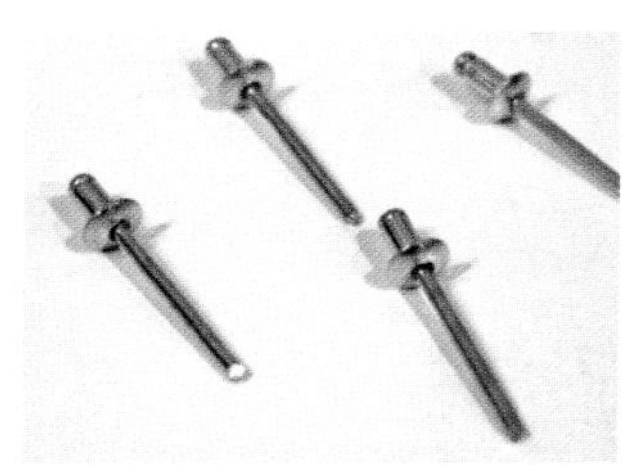
F

G

a) Name each of the fixings. *(3 marks)*

b) Name a suitable tool that would be used with each of the fixings. *(3 marks)*

10 Explain the advantages of using PVA glue. *(3 marks)*

11 Use notes and sketches to describe how to prepare and apply epoxy resin adhesive (Araldite) to a joint. *(8 marks)*

12 Use notes and sketches to describe how you would gloss paint a softwood door that contains knots. *(10 marks)*

UNIT ONE Written paper

GCSE

Design and market influences

In these chapters you will learn about:

- famous designers
- the natural world and how it influences design
- form and function
- market pull and technology push
- design periods through history
- how to be creative through sketching, modelling and computer-aided designing
- social, moral and cultural influences on design
- sustainability and environmental issues and how they affect designing
- designing for maintenance
- the role of the client, designer and manufacturer in designing a product
- how to analyse existing products
- different ways to present your ideas
- how to evaluate your ideas
- how to plan for manufacturing a product.

A *The Aerocar. In the 1950s, the designer Moulton Taylor designed this flying car. It is still flying more than 50 years later. It is an example of a very original design*

Creativity

This subject expects you to be creative, imaginative and innovative in your designing. You can draw on a variety of sources to help you to be creative. The following pages show examples.

A manufacturer may achieve greater sales if the product they sell is innovative. Think how mobile phones are continually being improved and given new functions.

When you work on your controlled assessment task in Unit 2 Design and making practice, you will be able to access higher marks if you show creativity and innovation. Try lots of the techniques, because your inspiration may come from an unexpected source!

B *The lemon squeezer. Phillipe Starck showed that a functional object can also be an object of beauty*

Communicating your ideas

It is important that you can communicate your design ideas clearly. A professional designer will use a range of techniques to develop products. Different techniques are well suited to each stage of the design process. Quick 2D and 3D sketches are very effective for getting ideas down quickly. Models are useful for assessing how well a design works, along with testing and improving the design. Three-dimensional CAD software allows you to view the product from any angle, produce formal drawings quickly, and produce photo-realistic presentations.

Issues that need to be considered in designing

A designer must take many factors into consideration in producing ideas for a product. There is a moral responsibility to produce products that have a lower impact on the environment. This has to be considered for the whole life cycle of the product. Can the product be maintained? What will happen to it at the end of its life?

A designer also needs to consider the different needs of people who are going to use a product. Who is going to use it, where is it going to be used, what are their cultural needs?

A designer needs to be able to evaluate products already in the market place. Can they be improved upon? The designer also needs to be able to compare and evaluate their own ideas.

C *Trevor Baylis developed the Baygen wind-up radio so that people in remote parts of Africa, with no access to power or batteries, could listen to radio messages about the spread of AIDS. His innovative technology is now available in a range of products*

3 Inspiration and innovation

3.1 Famous designers

Throughout history there have been many famous designers who have influenced the products that we use today. These people are creative geniuses. We can gain inspiration by looking at their work, finding out about their lives, how they lived and what influenced them to become truly inspired designers.

Objectives

Have an understanding of the work of several famous designers.

Understand the impact that those designers have had on the form and function of everyday products.

Philippe Starck (18 January 1949 –)

As a child, Philippe enjoyed taking things apart and re-building them. He spent many hours sawing, cutting, sanding and gluing materials to turn his ideas into working products.

From these early beginnings Philippe has continued to look at the objects that surround us and has reworked them to provide us with products that are not only easier and smarter to use but are also fun, original and exiting to look at.

His influence can be seen in many, many areas of our world. The list of products that he has been involved with includes buildings, furniture, vehicles, clothing, luggage, lamps, kitchen equipment and even toothbrushes.

A Chair designed by Philippe Starck

B Chair designed by Charles Rennie Mackintosh

Charles Rennie Mackintosh (7 June 1868 – 10 December 1928)

Charles lived in Glasgow towards the end of the Industrial Revolution, when there was an increasing demand for mass-produced items. His influences came from Asian styling and the emerging '**modernist**' ideas.

On leaving school, Charles began his career by working for an architect and by 1889 he had developed his own, immediately recognisable, distinctive style of design.

His influences can be seen in interior design, furniture and jewellery.

Key terms

Modernist: ideas from a period of design history around the turn of the 20th century. Modernism replaced traditional, extravagant products with very usable, stylish, original and exciting designs.

Hugo Alvar Henrik Aalto (3 February 1898–11 May 1976)

Alvar studied architecture at the Helsinki University of Technology in Finland from 1916 to 1921. In 1923 he opened his first architectural office. His style is modernistic using natural materials, warm colours and undulating lines.

His influences can be seen in architecture, furniture, glassware and paintings.

In particular, students following the Resistant Materials course will be interested in his work with laminated bent-plywood furniture (a process that he invented in 1932).

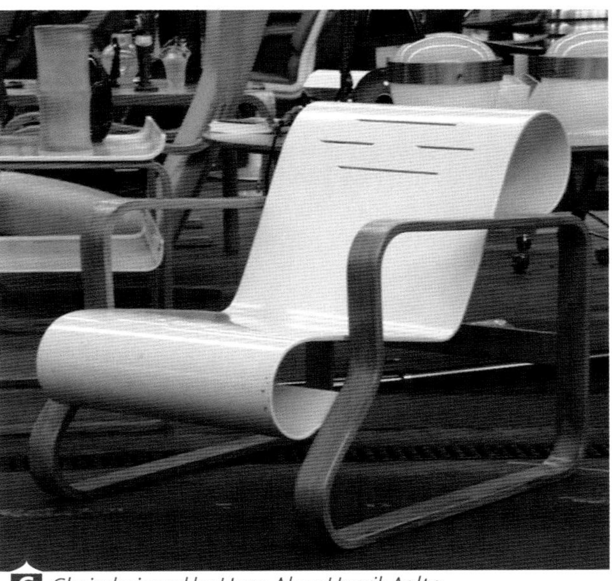

C *Chair designed by Hugo Alvar Henrik Aalto*

links

www.alvaraalto.fi
www.charlesrenniemac.co.uk
www.philippe-starck.com

Activities

1. Use the internet to research one of the following designers:

   Alvar Aalto | Philippe Starck | Marcel Breuer | Charles Rennie Mackintosh | Harry Ferguson | Sir Frank Whittle | Sir Norman Foster

   Produce a PowerPoint presentation that includes a brief biography of the designer and examples of their work.

   Deliver your presentation to the class.

2. Study the drawing of the simple table-top lamp (Diagram **D**).

   Redesign the lamp in the style of Charles Rennie Mackintosh.

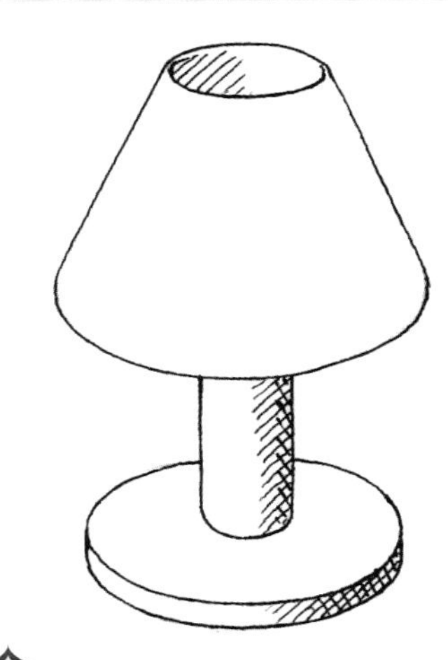

D *Simple table-top lamp*

Summary

Different designers have their own particular style.

Their styles can be used to develop new designs.

AQA Examiner's tip

- You should be able to recognise the work of several famous designers.
- You also need to know how to redesign a product in the style of a particular famous designer.

3.2 The natural world

The majority of design is concerned with making small adjustments and changes to existing products. This may involve altering the appearance of a product to make it look better (**form**) or by improving its performance to make it work better (**function**). However, occasionally, designers come up with something completely new and totally original. You can only do this by thinking 'outside the box', by looking at things that are not directly related to the problem that you are trying to solve.

Using the natural world as a source of inspiration is one method of improving your creativity.

Objectives

Understand the influences that objects found in the natural world have had on the form and function of everyday products.

Learn how to use natural objects as a source of inspiration.

links

See Form follows function on page 52.

Sea life – a giant clam

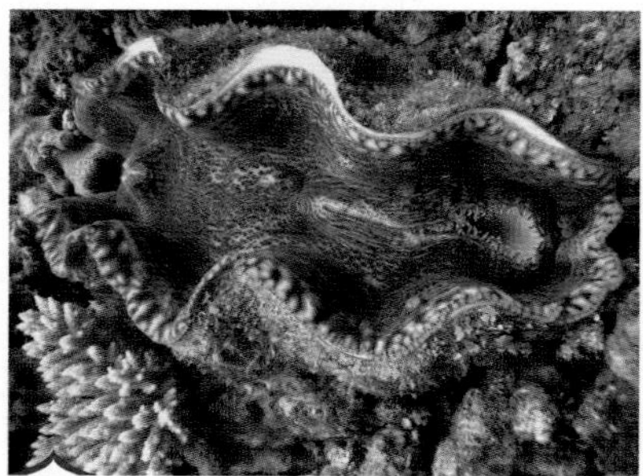

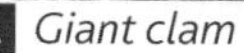

A *Giant clam*

B *Clamshell bucket*

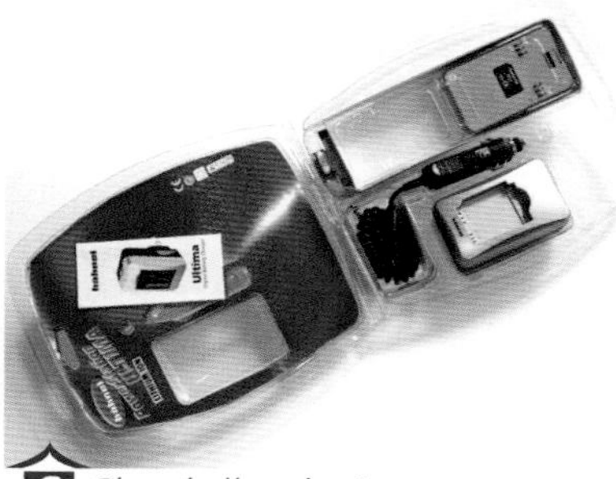

C *Clamshell packaging*

Take a look at the picture of the giant clam (Photo **A**). It has many natural features that have provided inspiration for designers of a number of very different products.

- The clam has a natural hinge that enables it to open and close.
- The bucket (Photo **B**) uses a fulcrum pin to allow the two sides of the bucket to pivot around.
- The packaging (Photo **C**) has a creased edge that gives the plastic flexibility to move around.
- The clam has a corrugated (ribbed) surface that gives it great strength.
- The bucket has a ribbed inside and a shaped outside to give it strength.
- The packaging has a domed and creased outside shape that provides it with strength.

Key terms

Form: form deals with the shape of a product.

Function: function deals with how a product works.

Fruit – an orange

An orange (Photo **D**) has many natural features that have been used by designers for inspiration.

The sphere has been used in many, many designs, not only for its low rolling resistance but also because of its strength.

D A peeled orange

E Sydney Opera House

Jørn Utzon's most famous work is the Sydney Opera House (Photo **E**). Incredibly, his design was inspired by the simple act of peeling an orange.

Human form – an arm

The human form is a highly sophisticated piece of design. You can take any part of the human body and you will find it contains many design features that have been used by designers as a source of inspiration.

Take a look at your own arm and compare it with the spot-welding robotic arm (Photo **F**):

- the robotic arm pivots in the same way that an arm does at the three key points of the shoulder, the elbow and the wrist
- each section of the arm has powered movement.

More ...

Other examples of the natural world influencing designers include animals, birds, insects, trees, flowers and rock formations.

Activity

1 Research one of the following reptiles on the internet. Use this research to produce designs for an interesting piece of jewellery:

a frog

b newt

c cobra.

Summary

Nature has been used as an inspirational source for some products.

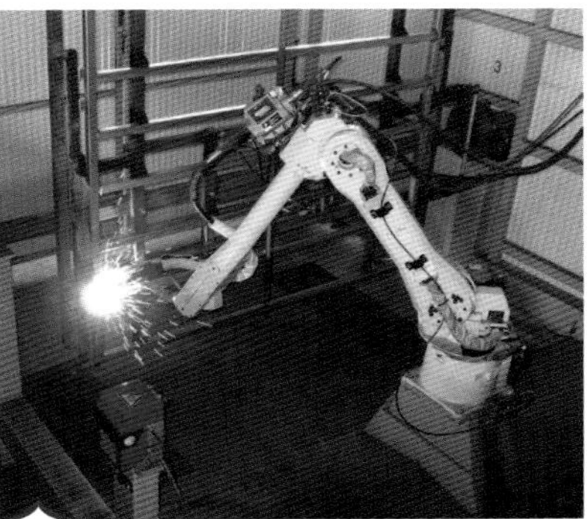
F Spot-welding robotic arm

AQA Examiner's tip

You should be able to produce clear, annotated and coloured 3D drawings that have been influenced by nature.

3.3 Form follows function

There are two main influences on a design. That is its form (**aesthetics**) and its function (what it should do). The two are often in conflict with each other. You may have a design idea that you like the look of but it does not work very well, or you may have a design idea that works well but does not look good. The secret is to get the balance right. Make sure the design works (function) then look at how you can shape it (form) to make it work and look better.

When designing a product for use by humans, a designer must make sure that it is the correct size for the group of people who will be using it, and that it is comfortable and easy to use. To achieve this, the designer could consult anthropometric data tables or carry out their own investigation with their specific user group.

Objectives

Understand the conflict between form and function.

Have knowledge of anthropometric data and how it is relevant to ergonomic design.

Key terms

Aesthetics: the features of a shape that make it look good.

Anthropometrics: measurements of the human body.

Anthropometric data

Anthropometrics is described as the measurement of the human body. Anthropometrical data provides us with the average sizes of the human body. The average size is known as the 50th percentile and is taken from the 5th to the 95th percentile (see Graph **A**).

If a designer is designing a games console that is suitable for use by teenagers and adults, then it is essential that they consult the relevant anthropometric table relating to measurements of the hand to ensure that it is the correct size for the intended user.

links

Look for other anthropometric tables on the internet.

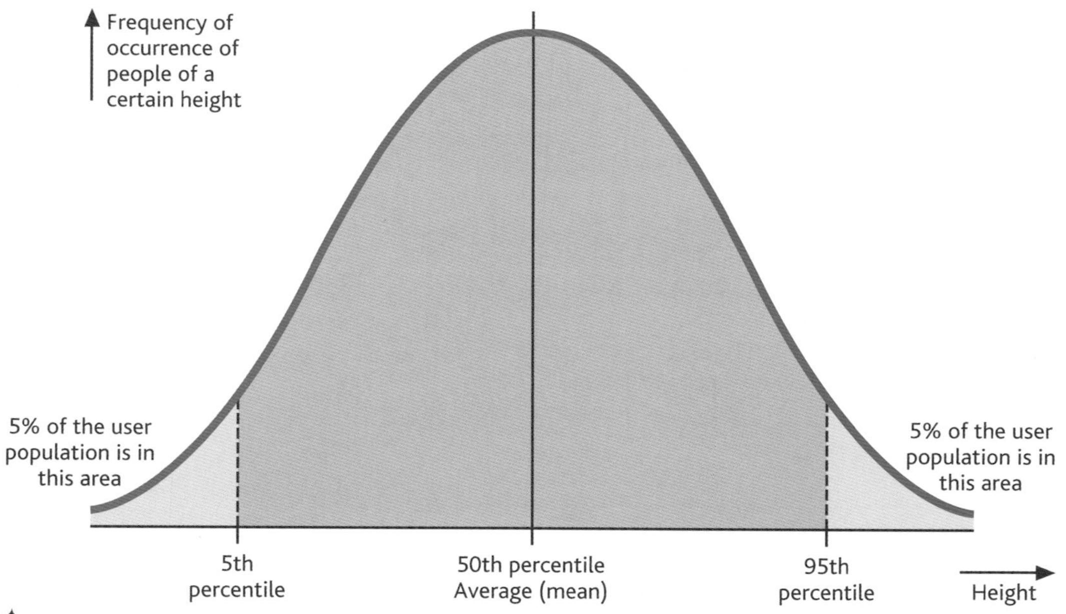

A *Anthropometric graph*

Ergonomics

Ergonomics is the study of people in relation to their working environment. It involves adapting tools, machines and general working conditions to fit the individual so that people can work at maximum efficiency.

A successful design must fit the body, be comfortable, safe and easy to use. Ergonomic features could include any of the following:

- Shape – the product may simply have rounded corners to aid safety or it could have a complicated shape designed to fit a particular part of the body.
- Texture – the product may have a ribbed or rubberised grip to prevent it slipping from your hand.
- Colour – the colour of the product may have a significant purpose. Under the bonnet of a car all the components that are required to be checked by the driver are normally clearly colour-coded for identification and safety purposes.
- Weight and size – if the product needs to be lifted or moved it will have to be manufactured to within a certain weight or size.

Case study

Cordless drill

When designing the cordless drill the designer will have to take anthropometric data into consideration to make sure that it is the correct size for the target group(s).

Here is a list of some of the ergonomic features found on the drill (Photo **B**):

- moulded handle to fit the hand
- rubberised grip on the handle
- second handle to aid stability
- key-less chuck
- balanced weight
- easy-read dial
- convenient trigger location
- quick-change battery
- convenient holder for a screwdriver bit.

B *Cordless drill*

Summary

The design of a product is influenced by form and function.

Anthropometric data is used by designers to ensure products are ergonomic.

Features that can be ergonomically designed include shape, texture, colour, weight and size.

Key terms

Ergonomics: the study of people in relation to their working environment, in order to maximise efficiency.

Activities

1. Find an example of a cordless toothbrush on the internet.

   Identify and describe its ergonomic features.

2. Study the drawing of the BBQ fork (Diagram **C**).

   Redesign the handle and clearly label its ergonomic features.

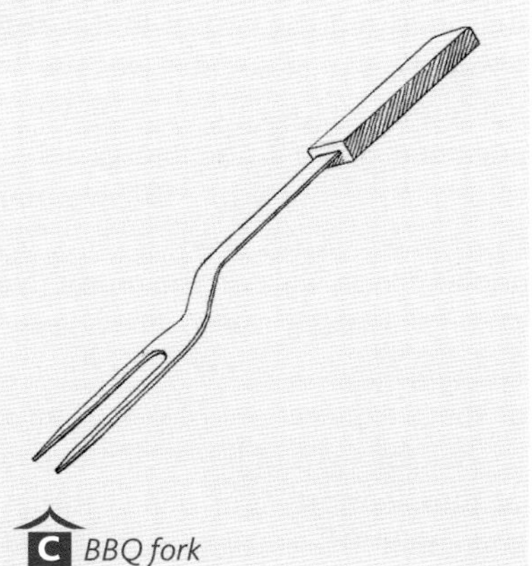

C *BBQ fork*

AQA Examiner's tip

- You should be able to describe clearly the terms anthropometrics and ergonomics.
- You should be able to identify and describe the ergonomic features of a product.
- You will need to produce designs that clearly display ergonomic features.

3.4 Market pull and technology push

The development of a design can be triggered in one of two ways. It can be the result of a social need for a product, known as '**market pull**', or it can be due to the result of advances in materials and manufacturing technology, known as '**technology push**'.

Objectives

Understand the terms market pull and technology push.

Understand how this influences the development of products.

Key terms

Market pull: the influence society has on the development of products.

Technology push: the influence new technologies have on the development of products.

Market pull

This is true problem solving. It begins with a demand or need for a certain product and it is up to the designers to become inspired and think of new and creative ways to solve the problem. The materials and manufacturing techniques that are required to make the product may not even exist, but designers are required to have the vision to dream up ideas that will solve the problem.

Case study

The 'space race'

Between 1957 and 1975 Russia and America began the exploration of space. This was known as the 'space race'. The technology to explore space did not yet fully exist but the two countries raced to ensure that their country was the first to put a man on the moon. Designers and engineers made it possible for satellites to be placed into orbit (1975), a man to fly in a space craft (Yuri Gagarin 1961), and a man to walk on the moon (Neil Alden Armstrong 1969). New designers and engineers have taken on the challenge of space exploraríon. The focus now is to look for signs of life within our solar system. Remote-controlled vehicles have been built, there has been a landing on Mars (Photo **A**) and humans are carrying out experiments in space. The same challenges are there for the designers and engineers of today and tomorrow.

A *Mars rover*

Activity

1. Make a list of all the products you can think of that have been developed as a result of 'market pull'.

links

See Chapter 6, Sustainability.

The manufacture and use of everyday products is causing environmental problems for our planet. We are using up our natural resources, polluting our atmosphere and harming our environment. This is of major concern to us all and has led to a huge market pull for more environmentally friendly products to be made. Designers and engineers have to come up with new materials, new ways of manufacturing and new products.

Case study

The hybrid car

The traditional petrol-fuelled car is a major polluter of our environment. There is a huge market demand for more environmentally friendly cars. The hybrid car (Photo **B**) is one of the answers to this problem. The hybrid car has an electric motor and runs on battery power as much as possible. When the batteries need charging or when extra power is needed a conventional petrol engine helps out. When going down hills the energy normally lost by braking is used to charge the batteries. When the car stops at traffic lights no engine is running.

B *Hybrid car*

Technology push

Most major manufacturing companies have research and development (R&D) departments, where engineers and designers work to improve materials and manufacturing technology. They then hope to use this new technology to improve their product range.

The mobile phone is one excellent example of the use of new technology. In 1980, when the phone first became 'mobile', it was simply a phone that could be moved around (Photo **C**). It had no extra features, had limited range and very large batteries that would quickly run down.

Over the years, advancements in technology have enabled the manufacturers to continue to develop the mobile phone. With each technological development comes a new and improved 'must have' feature.

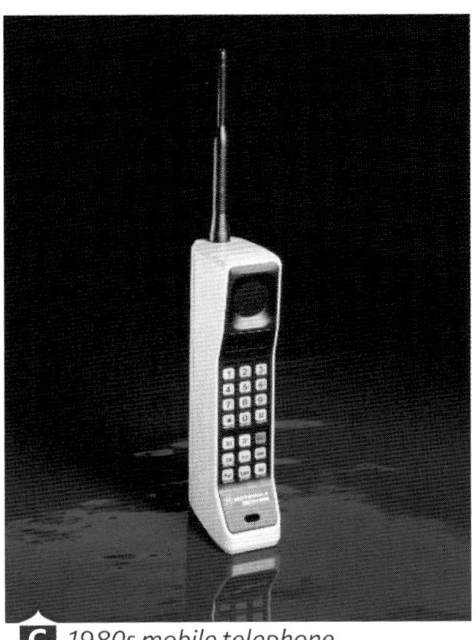

C *1980s mobile telephone*

Activity

2 Make a list of all the products you can think of that have been developed as a result of 'technology push'.

Summary

The development of a design can be caused by market pull or technology push.

Market pull is society's need for a product, such as the hybrid car, which meets the needs of people wanting to reduce pollution and damage to the environment.

Technology push is finding ways to use new technologies, such as in mobile phones.

AQA Examiner's tip

- You should be able to clearly describe the terms market pull and technology push.
- You should also be able to identify products that have been developed as a result of market pull and technology push.

3.5 Design periods through history

Throughout history there have been a number of key periods when there have been major changes in the styling of everyday products. These changes happened due to the arrival of new materials, new methods of manufacture, political influences, social influences, economic circumstances and the inspiration and vision of the designers and engineers of the time.

Objectives

Have a knowledge of the main design periods through history.

Pre-Industrial Revolution

Before the Industrial Revolution most people had very few possessions by today's standards. Everyday products tended to be hand-made on a small scale and of a poor quality. Only the very rich could afford to have many possessions, and products tended to be over-elaborate and expensive.

A *Arts and Crafts lamp*

The Industrial Revolution (late 1700s–early 1800s)

The Industrial Revolution allowed products to be manufactured in quantity for the first time. This often involved the use of machines and the division of the workforce into manufacturing cells. This meant that there was an increase in the number of products available. Products became affordable for more people to buy. Unfortunately the style and quality of these products tended to be very limited.

The Arts and Craft movement (1880–1910)

The Arts and Crafts movement aimed to place style and craftsmanship into everyday products. It was against what it viewed to be the dull and boring designs being produced by machines. Designers drew their inspiration from the use of natural materials and the use of organic shapes and natural forms.

Art Nouveau (1895–1905)

Art Nouveau (new art) aimed to place art into everyday products. Designers drew their inspiration from free-flowing lines, organic and natural shapes. The effect of this was that more intricate and elaborate designs were being produced.

B *Art Nouveau lamp*

Bauhaus (1919–1933)

The Bauhaus (building house) was a school of art and design that taught students the following design principles:

- Form follows function – design a product to work efficiently first, then see how it looks.
- Everyday objects for everyday people – designs should be for everyday use by everyone.
- Products for the machine age – products should be designed to be made using new materials and manufacturing processes.
- Simple geometric shapes – designs should use clean lines.

C Bauhaus lamp

Art Deco (1925–1939)

Art Deco was a very popular style that aimed to make everyday products stylish, elegant, functional and ultra-modern. Art Deco products feature geometric shapes, stepped forms and sweeping curves.

D Art Deco lamp

Utility (1940s)

During the Second World War materials needed to manufacture everyday products were in short supply. The manufacturing industry had been taken over by the demands of the war. This had a definite effect on the style of products that were being made. Designs had to be simple and functional as well as quick and easy to make from inexpensive materials. This was known as utility furniture.

E Utility chair

Post-war design

During the Second World War (1939–1945) there were many advances made in material and manufacturing technology. This had an influence on the everyday products that were made directly after the war.

Case study

The Maclaren buggy

During the Second World War, Owen Maclaren was a test pilot and a designer of the Spitfire undercarriage. A number of years after the war he observed his daughter struggling to manoeuvre his first grandchild in a conventional pram. As a true designer he used his knowledge of lightweight, collapsible structures to produce a solution. He invented the first Maclaren pushchair: a lightweight, easily manoeuvrable, collapsible pushchair. The chair has evolved over the years but the concept remains the same.

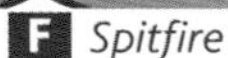

F *Spitfire*

G *First Maclaren buggy*

H *2009 Maclaren buggy*

Activities

1. Use the internet to investigate the style of products that are associated with the following design periods:
   a. Arts and Crafts
   b. Bauhaus
   c. 20th century.
2. Sketch a design for a dining room chair in the style of each of these three design periods.

Late 20th-century design

An ever-increasing demand for more sophisticated products combined with rapid advances in materials and manufacturing technology saw an explosion in the number of everyday products available to the consumer. This could be seen in the number of electronic products that were made available for all aspects of everyday life.

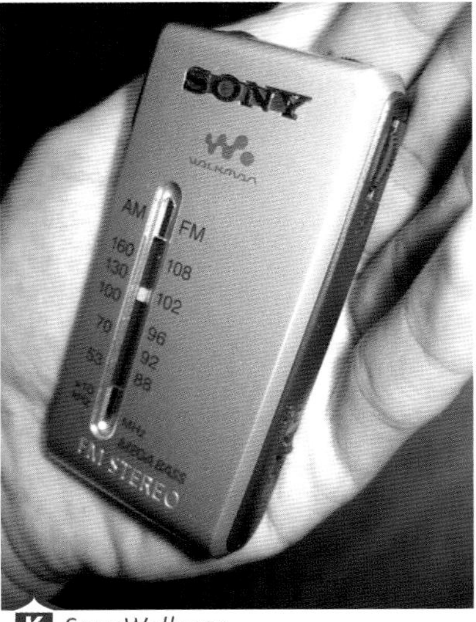

K *Sony Walkman*

I *Digital camera*

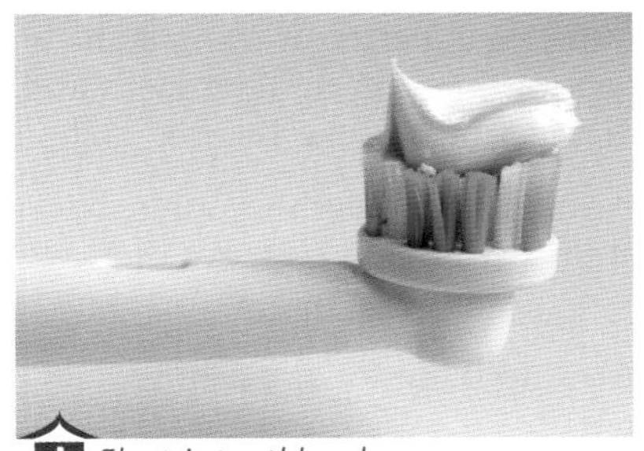

J *Electric toothbrush*

21st-century design and the future

A growing awareness of the environmental effects of using non-renewable resources to manufacture products is now having a major influence on the types of products that are being developed. Designers and manufacturers must take into account the full environmental impact of the products that they are developing.

An increase in the use of **nanotechnology** means that products are likely to become lighter, stronger and smarter.

An increase in the use of **microelectronics** means that products are likely to have even more design features.

Key terms

Nanotechnology: the technology used to rearrange individual atoms to create new, improved materials, systems and devices.

Microelectronics: the miniaturisation of electronics making products smaller and smarter.

Activity

3 Use the internet to investigate nanotechnology.

a Identify three different products that use nanotechnology.

b Explain how nanotechnology improves their performance.

AQA Examiner's tip

- You should be able to identify products from a certain design period.
- You should be able to produce clear, 3D, coloured, annotated drawings of products that have been clearly influenced by certain design periods.

Summary

There have been a number of key design periods through history that have had a major influence on the style of products that we use; pre-Industrial Revolution, Industrial Revolution, the Arts and Crafts movement, Art Nouveau, Bauhaus, Art Deco, utility, post-war, late 20th century, 21st century and the future.

links

www.nanotechproject.org

4 How to be creative

4.1 What is creativity?

Creativity is about thinking up new and exciting designs. It can involve simply changing part of a design to make it look better or to make it work better. However, real creativity is about thinking up new ideas and solutions to problems in ways that have never been done before.

We are probably all born with a certain amount of creative genius, but, as with most skills, we can practise, develop, train and improve our creative abilities.

It is about thinking 'outside the box', about looking for something different, something **unique** and original.

Objectives

Understand what is meant by creativity.

Key terms

Unique: something that is a one-off; there is nothing else the same as it.

Case study

'Concept' car

Most vehicle manufacturing companies will have a team of designers whose job it is to embrace and think beyond the limitations of new materials and manufacturing technologies.

These people design 'concept' vehicles (Photo **A**). These are vehicles that give us a glimpse into the future of transportation. They do not go into production but show the creative ability of the company and produce great excitement when they are displayed at motor shows.

A *Concept car*

Mood board

Researching is always a good way to fill your brain with things that are linked with the problem that you are trying to solve. One way that you can easily do this is to produce a mood board (Picture **B**).

A mood board is a single sheet of paper that has lots of images relating to your client and the problem you are trying to solve. Typical images that it may show are your client's taste in fashion, music, sport, and hobbies. Do not, however, include any images of existing solutions to the design problem. This would hold back your creative abilities.

B *Example of a mood board*

Thought shower

For some people, writing down their thoughts can be a source of inspiration. It usually starts with the problem labelled in the middle of a piece of paper, and then words linked with the problem and the client radiating out from the centre (Diagram **C**). The following acronym can help you to organise your thoughts:

ACCESS FM

- **A** – Aesthetics
- **C** – Cost
- **C** – Client
- **E** – Environment
- **S** – Safety
- **S** – Size
- **F** – Function
- **M** – Materials

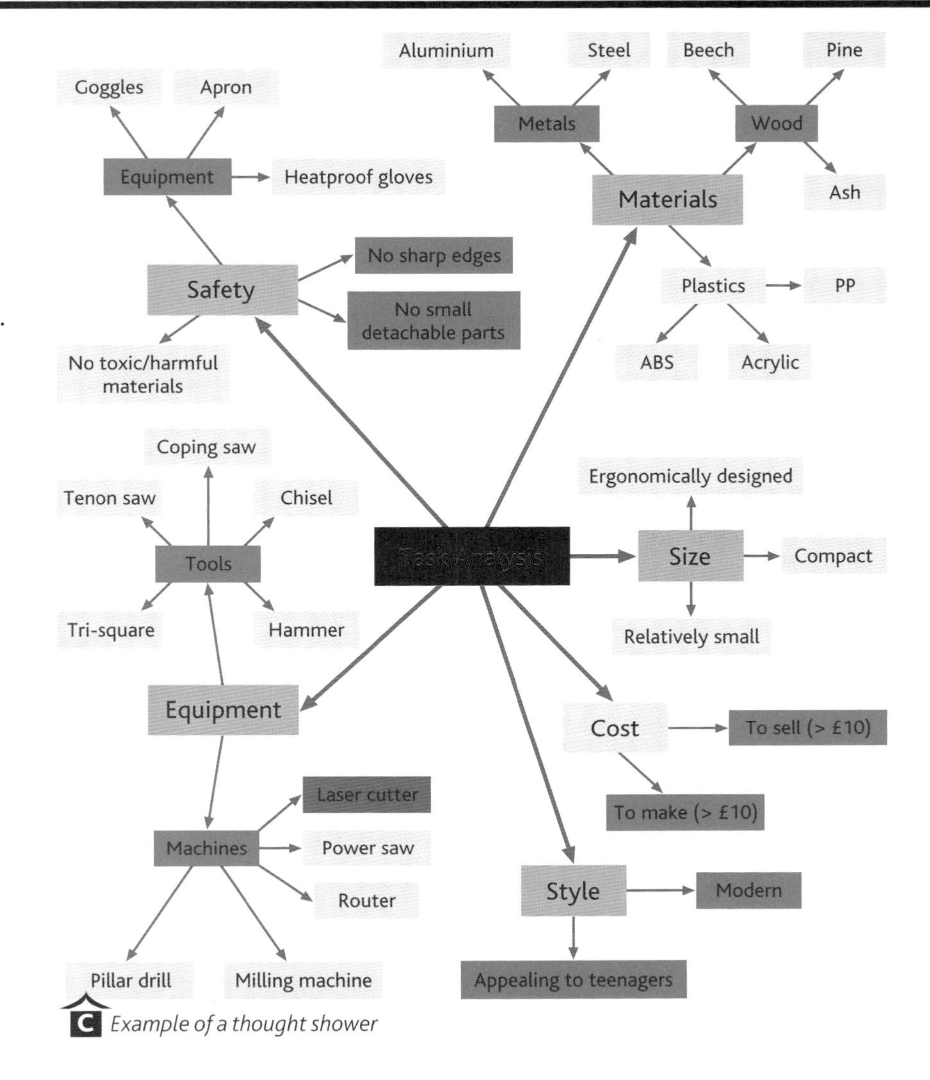

C *Example of a thought shower*

Incubation period

It is difficult to say how long it takes to come up with a new, completely original, exciting design. It could happen immediately or it may never happen. You can increase your chances of success by allowing yourself some time for the problem to 'roll around' in your head. This is known as the incubation period.

Pin your mood board and your thought shower up on the wall. As you think of possible designs, sketch them down, even if they seem unworkable and totally insane. Keep these sketches; this is real design thinking taking place.

Activity

You have been asked to design an iPod speaker system for use by a teenager.

Produce a mood board that represents your client.

Summary

People have to take risks and think beyond what exists at present to be creative.

Mood boards and thoughts showers are two methods of developing creative ability.

It is important to have an incubation period for creative ideas to develop.

AQA Examiner's tip

- Let your imagination flow.
- Do not be conservative in your designing.
- Log all your design ideas.
- Think 'outside the box'.

4.2 Developing your creative ability

Scruffiti

Producing original shapes can be quite difficult. We tend to think and design in box-like shapes. This is possibly because we find them easier to draw. For an original, non-box-like shape, try using 'scruffiti' (Diagram **A**). Take a piece of paper and very lightly produce random, curvy lines all over the page. Now look into your scruffy mess. Find interesting and original shapes. Line them in with your pencil and then add colour. Aim to produce as many different shapes as you can.

Objectives

Have an understanding of several different ways of developing your 2D creative ability.

Activity

1. Produce a range of original designs for a **pendant** using the 'scruffiti' method of creative designing.

A *Scruffiti*

Jack straws

'Jack straws' uses the same method as scruffiti but instead of drawing curvy lines draw straight lines (no need to use a ruler).

Geometric shapes

More shapes! Take a piece of paper and very lightly produce simple, overlapping circles, squares, rectangles and triangles (no need to use a ruler or a compass). Now look out for those interesting shapes. Line them in with your pencil and then add colour (Diagram **B**). Make it a contest between you and the person sitting next to you to find as many shapes as possible.

Key terms

Pendant: a piece of jewellery, usually worn hanging from a chain around the neck.

B *Geometric shapes*

Quick sketching

Sometimes you can force ideas out from your head and onto the paper. Try this one. Take a piece of paper and a black pen. Place the pen onto the paper and begin to draw your ideas. The secret is that you must not let your pen move off the surface of the paper. The theory is that as you begin to draw the part of your brain that stops your creative ability (the rational bit) is concentrating on keeping the pen on the paper, which allows your creative side to run riot!

Morphing

Morphing is the transition of one shape as it slowly changes into something else. It is a process developed for use in the film industry to depict characters changing into someone or something else. (Think of the film *The Incredible Hulk*.)

This process can be used to produce novel and amusing designs.

The illustration starts off with a mobile phone and a banana. If we produce an intermediate drawing (see Photo **C**) to show how it would

Activity

2. Produce a range of designs for a wall clock using the 'quick sketch' method of creative designing.

look half-way through the change we have produced a novel design of a phone that looks like a banana! Or is it a banana that looks like a phone?

Case study

Morphing

Here the designer has taken a regular mobile phone and a banana and morphed them together to produce an amusing and unique design for a novel mobile phone that looks like a banana. Or is it a banana that looks like a phone?

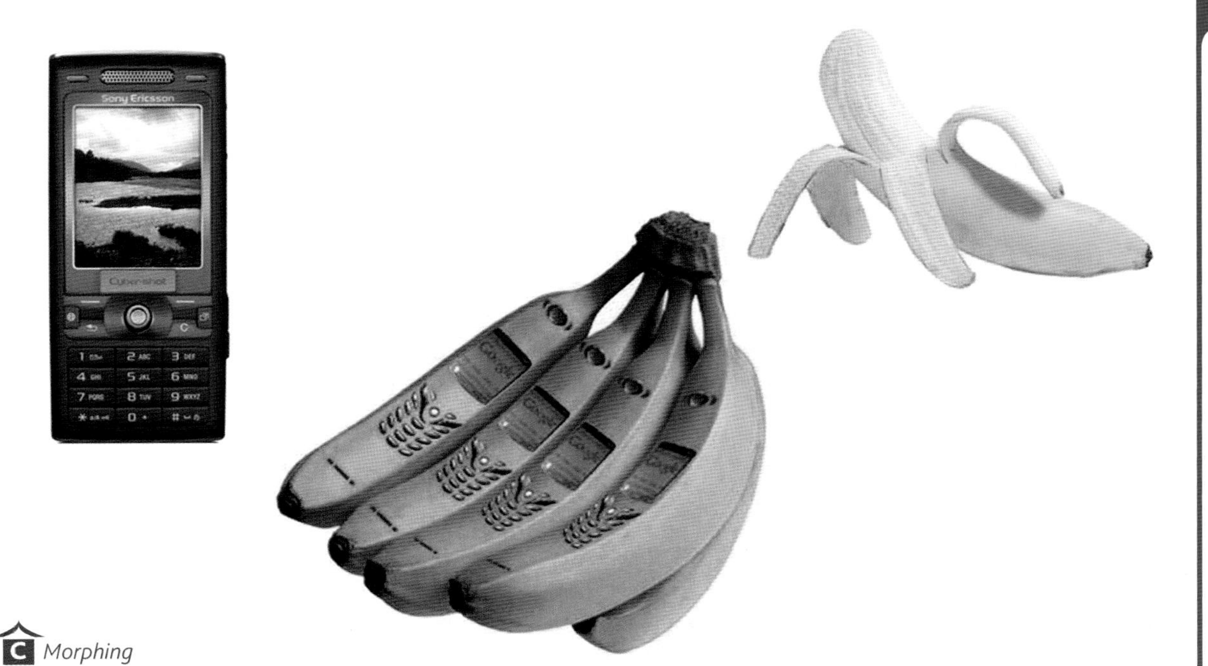

C *Morphing*

Activity

3 Find a picture of an apple on the internet.

Sketch out the chair you are sitting on.

Morph the two together to produce a new and different design for a chair.

Summary

Scruffiti, Jack straws, geometric shapes, quick sketching and morphing are all methods of developing creative ability.

links

Look at examples of these techniques in the controlled assessment section on page 146.

AQA Examiner's tip

- Try out as many methods of improving your creativity as possible and find out which ones work for you.
- Share your ideas with others.
- Get help and advice on developing creative ability from as many people as possible.

4.3 Using quick modelling and CAD

Quick modelling

So far we have looked at producing ideas simply in 2D form. It can, however, be very useful at the beginning of your designing to produce a series of quick models. No sketching required, just produce a 3D model straight away.

The first Dyson vacuum cleaner took James Dyson five years and 5,127 **prototypes** to develop before he eventually arrived at his final solution.

To create prototypes we need inexpensive or reusable modelling materials that are very quick and easy to work with.

Sheets of cardboard, cardboard boxes, cardboard tubes, paper, tape, scissors and glue are all cheap and readily available. You can use these to produce a model quickly. Remember, don't begin with a fixed idea in your head. Allow your brain to be creative and to come up with ideas as you are modelling.

Objectives

Learn different ways of developing 3D creative ability.

Key terms

Prototype: a model of a product that is used to test a design before it goes into production.

Ergonomic: something that has been designed to allow people to work efficiently by making it comfortable and user-friendly.

links

www.jamesdysonfoundation.com

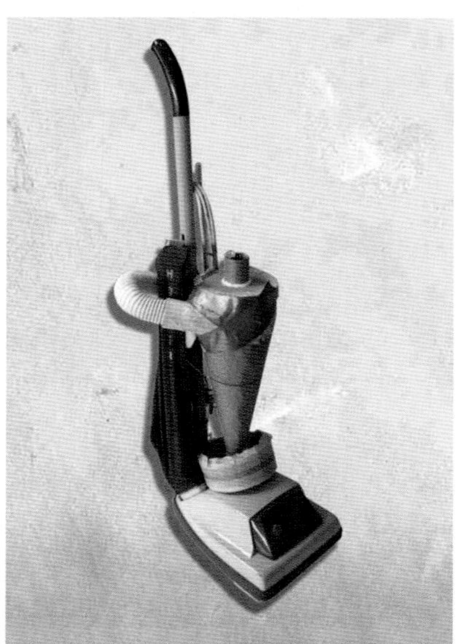

A *James Dyson's DC01 vacuum cleaner and a prototype vacuum model*

Commercial construction kits/modelling materials

Commercially produced construction kits are a useful tool that will enable you to produce quick 3D shapes for your design. Construction kits such as Lego and Meccano can be put together to produce simple shapes to help you visualise your design. The disadvantages are that the shapes you can produce are constrained to the regular forms of the parts of the construction kit.

Commercially produced modelling materials such as Plasticine offer a quick method of producing small 3D forms. It is excellent for producing **ergonomic** handles because of its highly mouldable nature. Its main disadvantage is that it does not set; however, it is reusable forever.

CAD (computer-aided design)

So far we have looked at quick designing methods that involve the use of traditional materials. However, CAD is a very powerful design tool. Programs such as 'Google SketchUp' and 'Techsoft 2D Design' are relatively simple CAD programmes that can be used to produce virtual 2D and 3D shapes quickly. These programs allow you to design 'on screen' so that you can produce and reject ideas very quickly. They allow you to edit your design in numerous ways. You can view it from any angle, render it any colour you want and even view how it would look if made out of different materials.

Some programs have a library of parts and components that you can download and use as part of your design.

CAD programs such as 'Pro-Desktop' and 'SolidWorks' are more sophisticated and will take your initial design further and help you develop it into a workable solution.

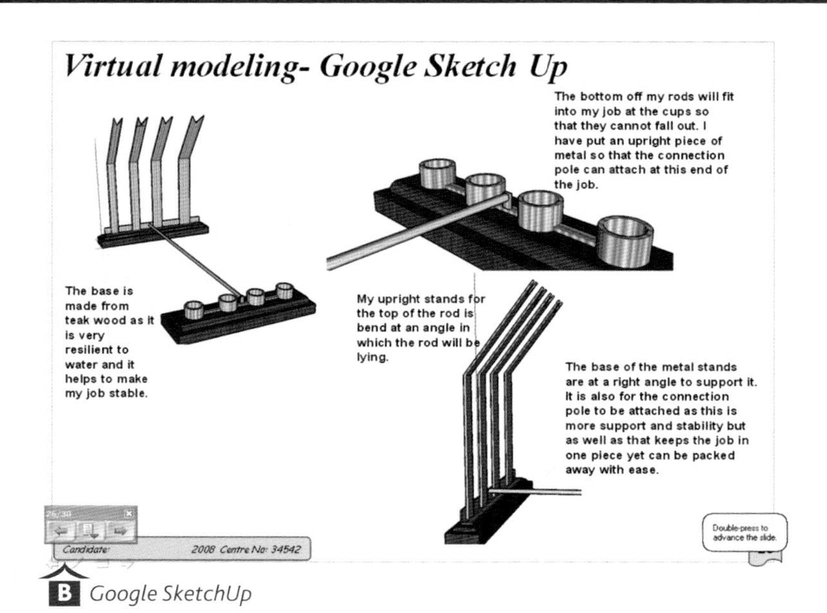

B *Google SketchUp*

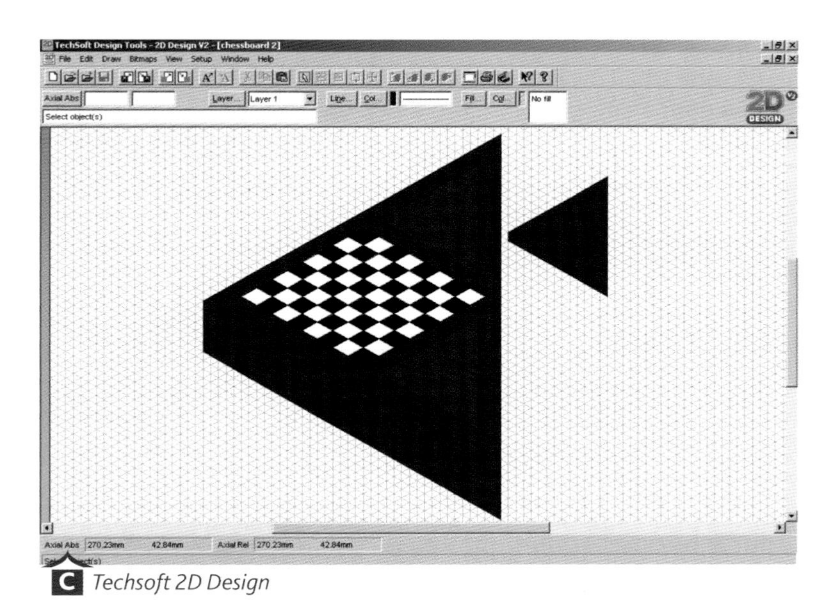

C *Techsoft 2D Design*

Activity

Google SketchUp is free to download onto your home computer. It comes with free tutorials to teach you how to use it.

Open Google SketchUp and produce a series of designs for a desk tidy.

Add images of stationery equipment from the Google SketchUp library.

Summary

Quick modelling, construction kits, modelling materials and CAD programs are several methods of increasing 3D creative ability.

links

See Evaluating design ideas, page 96.

AQA Examiner's tip

- Try out as many methods as possible to find out which ones work for you.
- Get help and advice from as many people as possible.

5 Design influences

5.1 Social and cultural influences

In the developed world, basic human needs are catered for very well. We have food, shelter, safety, education, freedom and lots of opportunity to interact with other people.

Objectives

Understand what effect society and culture has on the development of products.

Understand the effects that the developments of certain products can have on society.

Social influences

Society now demands that products are designed and manufactured to improve the quality of our lives and the environment in which we live. This stimulates an ever-increasing demand for newer, better, faster, smarter, greener products.

Example 1: Young people view mobile personal technology as a 'must have' item. This has led to rapid developments in products such as the mobile phone and personal hi-fi. Each new version includes another feature that ensures the last version will be seen as being **obsolete**.

Example 2: People are aware of the environmental effects of the manufacture and use of many of the products that we have in our lives. This has led to a demand by society that the products we use are now sustainable.

links

See Sustainability of materials on page 26.

Inclusive design

It is very important that all members of our society are taken into consideration when we are designing and making products. This is known as **inclusive design**.

Key terms

Obsolete: out of date, no longer required.

Inclusive design: designs that are accessible by all members of society.

Case study

Elderly people

Due to advances in health care we now have a society that is made up of more and more elderly people. It is essential that we consider their particular needs when we are designing products.

Disabled people

Disabled people have a right to access as many things in life as able-bodied people. Designers must ensure that their products are as accessible as possible to disabled people.

The wrist action tap makes accessing water much easier for elderly or disabled people (Photo **A**).

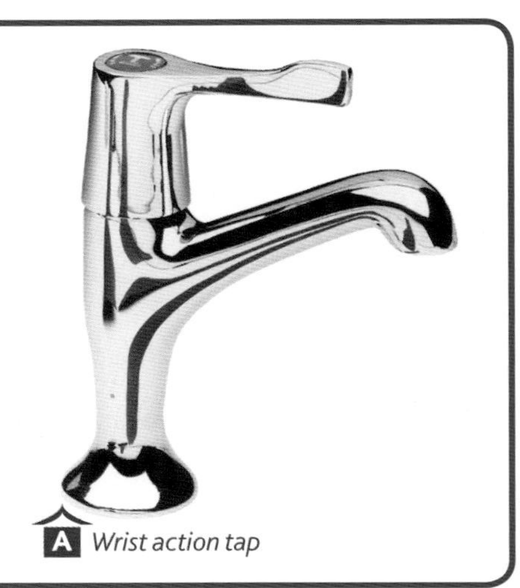

A *Wrist action tap*

Exclusive design

Some groups that make up our society have very specific needs and designing is focused on these needs. This is known as **exclusive design**.

Young children

Case study

The needs of young children and babies are very specific to them. Their needs are the main focus of the design.

The child's car seat is made exclusively for a young child to sit in (Photo **B**).

B *Child's car seat*

Key terms

Exclusive design: designs that are accessible only by particular members of society.

Culture: the beliefs, fashion, music, likes and dislikes of certain groups within society.

Cultural influences

Culture is the way particular groups of people have developed their beliefs, fashion, music and lifestyles. Our modern society is made up of people from a wide variety of cultures and we should use this to enrich our lives.

The cruiser motorbike reflects the 'laid back', 'easy going' image of the cruiser motorcyclist (Photo **C**).

C *Cruiser motorbike*

Activity

1 Search the internet or magazines for images of the following different types of car:

a sports car

b family car

c off-road vehicle.

Identify and explain the features that make these cars unique.

Summary

Society and different cultures can influence the design of a product.

AQA Examiner's tip

- You should be able to analyse a product and evaluate how it has been influenced by society.
- You should also be able to analyse a product and evaluate how it has been influenced by a particular culture.

5.2 Moral implications

Designers and manufacturers have a responsibility to society to make sure that the products they design and make are morally correct. This means that the products they produce should not harm or offend anyone, in any way. Moral decisions are often a matter of asking ourselves questions about the rights and wrongs of a product.

Objectives

Understand that there are moral implications to be considered when designing, manufacturing and using products.

Designers

Typical questions that designers must ask themselves are:

- Will the design offend any particular group of people?
- Does the design exclude certain groups of people?
- Will the design harm anyone?
- Will the design have a bad influence on certain people?

Manufacturers

Typical questions that manufacturers must ask themselves are:

- Is the product sustainable?
- Does the manufacture of the product have any negative effects on the environment? (See Photo **A**).
- Are the workers treated in a fair way?
- Are materials being used efficiently and effectively?

Some products now display the FAIRTRADE Mark. This shows the customer that the product has been certified by the Fairtrade Foundation, which believes in a better deal for producers in the developing world.

A *Pollution*

Users

Users of products have a very big influence on the types of products that designers and manufacturers produce. As users we should ask ourselves questions about the morality of the products that we buy.

Typical questions that users must ask themselves are:

- Will the product that I am going to buy offend, harm or have a negative influence on anyone?
- Is it made from sustainable materials?
- Has its manufacture had any negative effects on the environment?
- Have the workers been treated in a fair way during its manufacture or has the product been made in a **sweatshop**? (See Photo **C**).

B *No Sweatshops*

C *Sweatshops*

Case study

The motor car

The motor car is a part of our everyday lives. It gives us the freedom to transport people where and when we want.

Before buying a car, a motorist should examine the **moral issues** concerning their purchase.

Moral decision 1: Do I really need a car?

- Would walking, cycling or using public transport be a better option?

Moral decision 2: How safe is the car?

- Does it have crumple zones? – Areas of the car that collapse in a crash, absorbing the impact and helping to protect the people inside the car.
- Is it fitted with air bags? – Bags that rapidly inflate with air in a crash.
- Does it use anti-locking brakes? – Brakes that stop the wheels from locking under hard braking, preventing the car from skidding off the road.

Moral decision 3: Is the car as environmentally friendly as possible?

- How energy efficient is it?
- How much of the car is recyclable?

Key terms

Sweatshop: a factory where people work for long hours, in poor conditions and for little pay.

Moral issues: the right or wrong of an action.

links

See Sustainability of materials on page 26.

Activity

1 Evaluate the moral issues concerned with each of these products:

a aircraft
b composting bin
c bicycle.

Summary

The moral issues of a product's design, manufacture and use must be considered.

AQA Examiner's tip

You must be able to analyse a variety of products and describe the moral implications of their design, manufacture and use.

6 Sustainability

6.1 Sustainability and environmental issues

Every product we use has an impact on the environment. The materials they are made from have to be extracted, transported, processed and manufactured. The products are transported to the consumer and then need to be dealt with at the end of their life.

Each stage of this life cycle of a product has an impact on the environment. New laws are frequently being introduced that make designers and manufacturers increasingly consider the environment. Many consumers are making choices about the products they buy and use based on environmental factors.

Objectives

Understand that designers and consumers must consider the sustainability and environmental impact of products.

Products that can be reused

One choice the consumer can make is whether to buy products that can only be used once, or those that can be used over and over again. A plastic milk bottle (Photo **A**) is used once and then it is disposed of; however, a glass milk bottle (Photo **B**) can be reused many times.

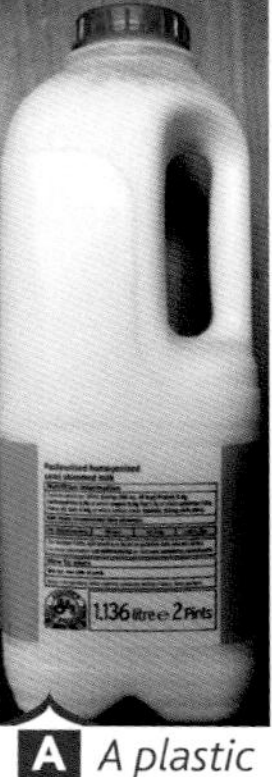

A *A plastic milk bottle*

B *A glass milk bottle*

Products with parts that can be replaced

Some products are designed so that only part of them is disposed of after use. This avoids having to replace the whole product. To improve

Case study

The razor

The design of razors has become less sustainable over the last century, as an increasing number of parts are **disposable**.

- The original 'cut throat' razor had a large blade that could be sharpened over and over again. This was a sustainable product because once you had bought one, it could last a lifetime.
- In 1903, Gillette launched the first safety razor. It had a separate handle, blade holder and disposable blade. Once a person had bought the razor, they would be obliged to keep buying new blades. This made the product less sustainable than the cut throat razor; it also made the manufacturer more money!
- Modern razors have a cartridge head that is disposed of after use. The cartridge contains several metal blades; these are held in plastic. This means that more materials are disposed of, making the product less sustainable than the safety razor.
- Some razors are now completely disposable. When the blade is blunt, the whole product is thrown away.

C *Four types of razor*

the **sustainability** of a product, designers should look to decrease the number of parts that have to be disposed of after a short period of use.

Other sustainability issues

The designer, manufacturer and consumer all need to consider the environmental impact of the materials used and the processing of those materials to make a product:

- Are the materials from sustainable sources?
- Are huge amounts of energy needed to process them, and transport the product to the consumer?
- What happens to the materials in the product at the end of its life?

Case study

The supermarket carrier bag

Until recently, most supermarkets in the UK issued unlimited plastic carrier bags to customers. These carrier bags, made from oil-based plastics, were designed to be disposable. Oil is a limited resource that will run out one day. When sent to **landfill**, the carrier bags could take hundreds of years to break down.

In response to these problems, supermarkets introduced a number of initiatives:

- collection of used carrier bags to allow them to be recycled
- reward points on customer loyalty cards for those reusing old carrier bags
- introduction of **biodegradable** bags (made from plant products and able to be composted in less than two years)
- 'a bag for life' – more sturdy bags that can be reused many times, and that would be replaced free of charge at the end of their usefulness
- 'no carrier bag' option for home delivery, with a loyalty point bonus as an incentive.

Each of these steps has increased the sustainability of the plastic carrier bag.

Recently, some supermarkets have introduced cotton carrier bags. These are even more sustainable because:

- the raw material for the bag is from a renewable source – a plant
- the cotton is often processed close to where it is grown, reducing transport energy/pollution
- the bag is much stronger than a plastic carrier bag, so it lasts for much longer
- at the end of its life the bag can biodegrade on a compost heap.

Activity

Why do some people like to change their mobile phone often?

List some steps a designer could take to encourage people to keep their mobile phone longer?

Key terms

Disposable: a product that is designed to be used a limited number of times before being thrown away.

Sustainability: the ability to keep making or using a product without excessive damage to the environment.

Landfill: a large hole in the ground that is filled with rubbish that is not being recycled. Sometimes old quarries are used for this.

Biodegradable: a material that breaks down naturally with time. Sunlight, rain or bacteria could break down the material.

links

See Sustainability of materials on page 26.

D *Cotton bags are from a renewable material, can be used many times and will biodegrade. Plastic bags are designed to be used once and then be disposed of*

AQA Examiner's tip

Consider the complete life cycle of products, from raw material, processing, manufacturing and use through to disposal at the end of use.

Summary

Consumers have choices about the sustainability of products they buy.

Designers can take steps to make products more sustainable.

kerboodle!

6.2 Designers, manufacturers and product sustainability

The mobile phone: disposable or sustainable?

Many manufacturers are accused of encouraging people to dispose of products, which work perfectly, in order to buy a new product. The mobile phone is a good example of a product with built-in **obsolescence**. Many people buy a new phone every 12 or 18 months, because a 'better' phone is available. Their old phone may still work perfectly, but it is no longer desirable to them, and is often thrown away. The appeal of a new phone may be:

- a smaller casing
- more functions
- a bigger memory
- a better camera.

This constant upgrading of mobile phones makes the product less sustainable. It is likely to be bought, used for a relatively short period, and then disposed of.

Consumer choice

In order to make the mobile phone more sustainable as a product, designers could design the phones to be upgraded rather than replaced. This would encourage people to keep them longer.

- The software on some mobiles can already be updated.
- The casing of some phones can already be replaced to keep the phone looking fashionable.
- The camera element could be a slot-in unit that could be replaced as more powerful cameras become available.
- The memory card slot could be standardised between different phone models.

Designer/manufacturer responsibility

The mobile phone could also be made more sustainable by:

- using renewable materials for the casing (vegetable-based plastics)
- using fewer materials in the product
- using materials that require less energy in manufacture
- designing more energy efficient circuits
- designing mains chargers that switch off when the phone is fully charged
- offering alternative charging methods such as solar power and hand-cranked chargers
- reducing the packaging of phones, and using recycled materials for the packaging
- designing with classic styling that remains fashionable
- enabling phones to be repairable
- improving recycling opportunities for the end of the life of the mobile phone.

Objectives

To understand that designers have a responsibility to create products that are more sustainable and have less impact on the environment throughout their life cycle.

Key terms

Obsolescence: lack of appeal to consumers because something goes out of date and better products become available.

A *This phone is many years old but still works perfectly*

Activities

1. Work in small groups to assess the functions of a mobile phone. (You will need one mobile phone to review per group.)
2. In a table, list all of the mobile phone's functions.
3. For each function, state how often you use that function.
4. List the features you would look for in a new phone.

The law and recycling

The European Union (EU) has been introducing new laws to insist that manufacturers and consumers consider the environment when a product comes to the end of its life.

End of Life Vehicle Directive (ELVD)

The ELVD restricts the use of toxic materials in new vehicles. All plastic parts have to be labelled to aid recycling. Vehicle manufacturers have to publish information about how to dismantle the vehicle.

Waste Electrical and Electronic Equipment Directive (WEEE)

Similar to the vehicle directive, this encourages consumers to take their old devices to WEEE collection points. Manufacturers have to arrange collection from these points. Designers have to make the products easier to dismantle, reuse and recycle.

Energy Labelling Directive

All electrical items such as washing machines and refrigerators have a label on them describing their energy efficiency rating (Photo **B**). This encourages consumers to make choices that are better for the environment.

There are more laws and directives to cover other products, packaging and eco-labelling.

Making products easier to recycle

Smart materials (materials that have a reactive capability) can be used to make products easier to take apart at the end of their lives. Shape memory polymers and shape memory alloys are materials that can be designed to change shape at specific temperatures. By replacing screws, rivets and clips with fastenings made from smart materials, a product can be designed to 'fall apart', with different materials releasing at specific temperatures (Photo **C**). This process is known as **active disassembly**.

Making new products from recycled materials

When products are recycled at the end of their lives, it often produces a brand new material. Designers and manufacturers have to work together to invent new uses for these new materials. Examples of this include:

- plastic bottles being used to make clothing (fleeces)
- old car tyres ground up and re-bonded to make flooring for children's outdoor playgrounds.

Summary

Common products are being developed to improve their sustainability at all stages in their life cycle.

links

See Smart and modern materials on page 24.

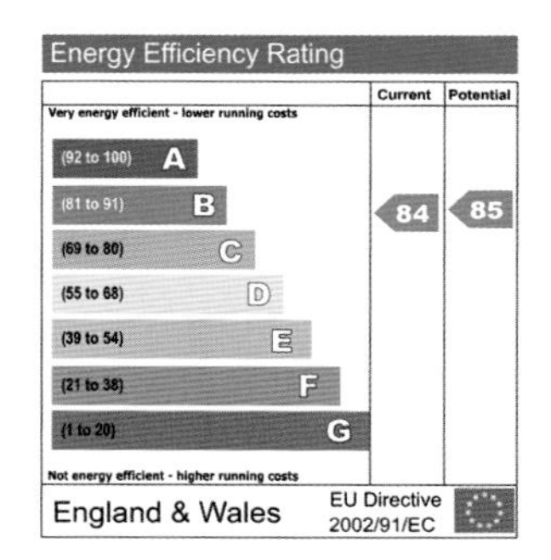

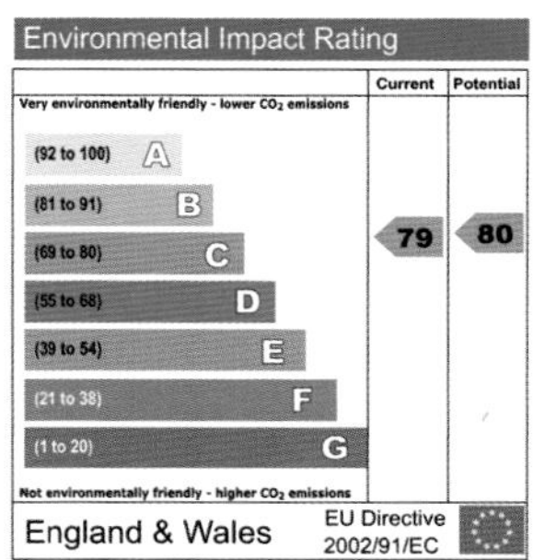

B *Energy rating label from an electrical product*

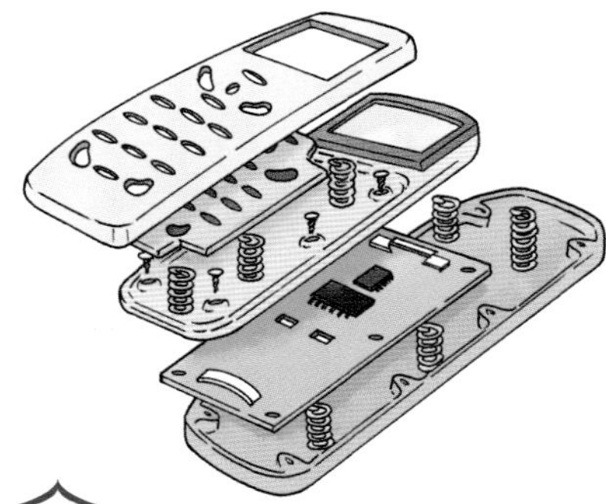

C *A mobile phone coming apart with smart material fastenings*

Key terms

Active disassembly: taking apart a product semi-automatically, by using smart materials for the fastenings. The fastenings release each part of the product at a pre-set temperature.

AQA Examiner's tip

Sustainability of products will be a key feature of examination questions.

kerboodle!

6.3 The 6 Rs

Many consumers try to be 'green' and think of the environment and sustainability when they buy things. Designers and manufacturers are required by law to try to reduce the environmental impact of the products they create. Six key words summarise approaches that can be taken by the consumer, designer, manufacturer and retailer – the **6 Rs**:

- reduce
- recycle
- reuse
- refuse
- rethink
- repair.

Objectives

Understand the meaning of the words of the 6 Rs.

Be able to name and explain examples for each category.

Key terms

6 Rs: six words beginning with the letter R. Each word describes action that can be taken to reduce the environmental impact of products.

Reduce

Consumers need to look to reduce the number of products they buy, or consider buying products that use less energy. Manufacturers are looking to design products that:

- have fewer materials in the product
- take less energy to manufacture
- need less packaging during transport.

Retailers can reduce carbon emissions by transporting products straight to the consumer from the place of manufacture, instead of via warehouses and shops.

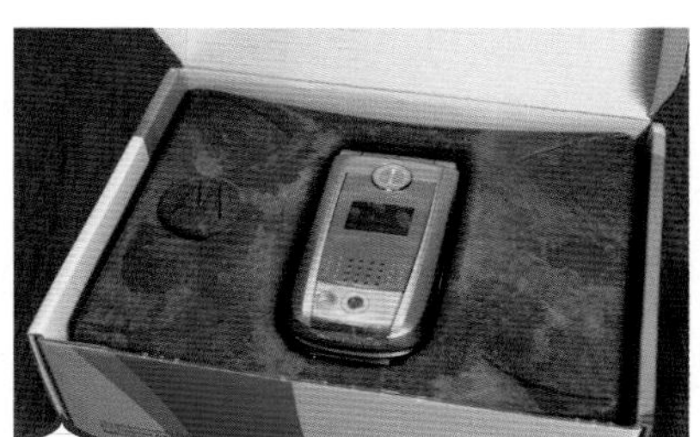

A *An example of an over-packaged product*

Recycle

Products are converted back to their basic materials and remade into new products, as with the following examples:

- glass crushed, melted and made into new bottles
- aluminium cans melted down to make new products
- plastic bottles recycled into drainage pipes and clothing. (It takes 25 x 2-litre plastic bottles to make one fleece jacket.)

Designers and manufacturers need to design products for recycling. Car manufacturers are obliged to label all plastic parts in new cars to aid recycling. Electrical and electronic products now have to include recycling instructions for the consumer. These state that products should be taken to special recycling points at the end of their life. Many local councils now collect recyclable materials separately from normal domestic waste. Products collected include paper, cardboard packaging, steel and aluminium food cans, plastic bottles, glass bottles and jars.

Americans throw away 25 million plastic bottles per hour

Plastic recycling firm reports shortage of plastic bottles

B *Newspaper headlines regularly report obstacles to recycling*

Activity

1. Look at plastic products around the home that display the recycling symbol. Has the manufacturer included a code in the symbol to name the plastic?

   Make a list of products found and write down the codes.

Reuse

Glass milk bottles are a classic product that is reused. A more recent product that can be reused is a printer cartridge, which can be refilled (Photo **C**). Some products have filters that can be washed rather than disposable, single-use filters. Most coffee machines have a single-use disposable paper filter. Some can use a metal or plastic filter that is able to be washed and reused repeatedly.

Consumers could sell or donate products they no longer use themselves so that someone else can use them. Some charities will collect items for reuse, such as the Woodland Trust, which collects old Christmas cards.

Designers need to consider how a product may be dismantled at the end of its life so that parts may be reused.

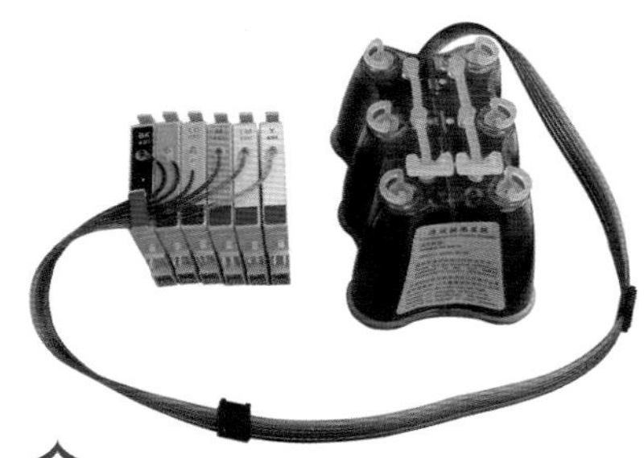

C *A printer cartridge being refilled to allow its reuse*

Refuse

The consumer has the choice as to whether they buy a product or not. They can ask the following questions:

- Should they refuse the product because it is too inefficient (in use, or in its use of materials)?
- Should they refuse the product because its packaging creates too much waste? (Disposable cups from a coffee shop, plastic carrier bags from a supermarket, plastic water bottles, etc.)

The designer and manufacturer have an increasing need to think about how the consumer will react to their products. Will the consumer refuse them?

D *A reusable cup is 'greener' than this disposable cup*

Rethink

Consumers can ask questions such as:

- Do I really need this product?
- Can I do things differently?
- Can this product be shared with another consumer?

Designers and manufacturers can make products that do the same job more efficiently. They can design the packaging so that it is easier to recycle (for example, by making the packaging from a single material).

Repair

Designers have a responsibility to design products that can be repaired more easily. It takes fewer resources to replace a part of a product than to replace the whole item.

Summary

Designers and manufacturers need to produce products that can be reused or recycled easily.

Products need to be efficient in their use of resources.

The consumer needs to decide if they really need the product.

The consumer is responsible for what happens to a product at the end of its life.

AQA Examiner's tip

- Understand the meaning of each of the 6 Rs.
- Be able to describe how the 6 Rs could be applied to everyday products.

Activity

2 Applying the 6 Rs to everyday products.

For each of these products, describe how the consumer or designer can think about the 6 Rs:

a mobile phone
b coated cardboard drinks carton (Tetrapak)
c wooden dining table
d metal desk lamp
e child's toy.

6.4 Designing for maintenance

Maintenance can prolong the life of many products. A dishwasher will need maintenance such as regular cleaning of the filter to remove food particles. Some dishwashers will also need salt and rinse aid added periodically.

Objectives

Understand that some products need to have maintenance carried out on them.

Understand that designers need to design products to allow for this maintenance.

Key terms

Maintenance: cleaning, adjustment, lubrication or replacement of parts of a product, to allow it to continue to function correctly.

Electronic devices

An electronic device may need the batteries changed. Some electronic devices, such as hand-held computers, may have main batteries that need changing regularly, and a back-up battery that would only need to be changed infrequently. The designer may design the cover for the main batteries so that it can be slid open without tools. It will also slide back and click into place. The cover over the back-up battery might be held in place with a screw, requiring the user to open it with a screwdriver. This is a deliberate design feature, forcing the user to think about the operation they are carrying out.

Parts the user should not take out, such as circuit boards, will be held in place with special screws. These would need to be undone with a specific screwdriver designed to fit only that type of screw. Only the manufacturer and authorised repairers would be likely to have these screwdrivers.

A *A dismantled mobile phone*

Mechanical devices

Devices with moving parts are likely to need regular maintenance. This may be lubrication, adjustment to allow for wear or replacement of parts. The designer will need to allow for this, particularly where moving parts are hidden for safety. It may be that a panel will need to be held on with screws or bolts (for example, the cover over the drive belts on a pillar drill in the school workshop).

As well as allowing access for maintenance, a designer will have to think about how adjustments will be carried out. A device driven by a belt will need either a tensioning device, or the device will need to be mounted on a bracket that can be tilted to increase the distance between the two pulleys (Picture **B**).

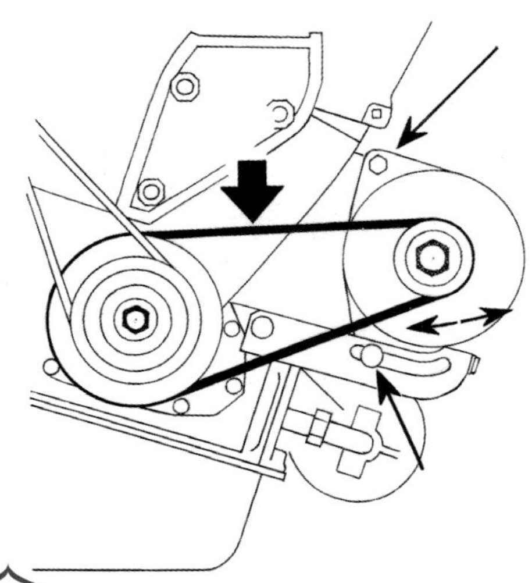

B *Adjusting a drive belt for a car alternator*

links

Look at the online resources for more examples.

The responsibility of the manufacturer

When manufacturers bring new products to the market, they need to produce maintenance information for both the consumer and for people whose job it is to repair these products. The repair manual is likely to be much more detailed. For the consumer, information will only relate to operations that the manufacturer intends them to carry out themselves. This information is often displayed in diagrams. Products intended for markets in many countries will also need multilingual instructions.

Reducing maintenance

Consumers often prefer to buy products that need less maintenance, or are maintenance free. Cars used to require a math range, or service, every 3,000 miles. Many now need servicing by a garage only every 20,000 miles. Car batteries used to need checking and topping up with distilled water regularly. Most are now 'sealed for life' items that need no maintenance. A mountain bike, on the other hand, needs regular maintenance if it is to work well and last a long time.

C *A mountain bike*

Activity

Look at Photo **C** of a mountain bike.

Make a table with three columns. In the first column, list the maintenance operations that a user would need to carry out. In the second, state whether the operation is cleaning, adjustment or replacement. In the third, describe how the designer has made the maintenance possible.

Summary

A designer needs to design products so that maintenance procedures can be carried out where these are necessary.

Some maintenance can be done by the consumer, but some is intended for authorised repairers only.

AQA Examiner's tip

Be aware of the different sorts of maintenance activities that need to be carried out on a range of products.

7 The client, designer and manufacturer

7.1 The client

A client is someone who asks a designer to design a new product. The product may be for them, or it may be for other people to buy and use (the user).

Objectives

Understand who a client and user are.

Understand the role of the client in the design process.

The bespoke item

A new fitted kitchen is an example of a bespoke item. It is a one-off design for that house only. The house owner (client) will invite a designer into their home to produce designs for the new fitted kitchen. The designer will measure the kitchen and assess the needs of the house owner. From what they learn on the visit, the designer will produce design ideas for the kitchen. In this example, the client and the user of the product are the same person.

A *A fitted kitchen for an individual house is a bespoke item*

Activity

Write a list of questions that a fitted kitchen designer might ask on a visit to a client.

Designing when the client and user are different

The case studies below show situations where the client and user are different. The client will instruct a designer to produce designs that meet the needs of the users.

links

Look at examples of market research in the students work on page 142.

Case study

New furniture for a hotel chain

A budget hotel chain with 200 hotels across the UK wishes to replace all its bedroom furniture. The new furniture will be used in every one of its 9,000 rooms, so that every room will look alike. The client is the hotel management. The user will be the people who stay overnight in the hotels.

B *Furniture for a hotel chain*

Case study

A furniture shop extending its range

An upmarket furniture shop wishes to extend its range of furniture to include bedroom furniture. The furniture shop is the client. They will approach a designer and commission them to design the furniture.

C *A luxury swimming pool*

D *A supercar*

These are the sorts of products that may be bought by the target market of the upmarket furniture shop. A designer needs to consider the styles of these upmarket products when designing for this target market.

Creating a consumer profile

The designer must think about who is going to use the furniture in these case studies.

- Who is the **target market** (target group) for the product?
- What are the needs of these users?
- What sorts of furniture do they already own?
- What sorts of other products do they use? What styles do these have?

The designer can use this information to build a picture of the typical user. This is called a user, or consumer, profile. Designers will use a range of techniques to create this profile. This process is called **market research.**

Client involvement in the design process

A successful designer will refer back to the client at each stage.

They will use the client to:

- help build a **user profile** (also known as a **client or consumer profile**)
- help identify shortcomings in existing products on the market
- consult about design ideas, sketches and prototype models.

Summary

Sometimes the client and the user may be different people.

A designer will carry out market research to find out the needs of users of a new product.

The designer can use information gathered to build a picture of a typical user of the product.

Key terms

Target market/target group: the people who, it is hoped, will buy and use a new product.

Market research: collection of information from likely users of a product, to find their needs and what they would like.

User/client/consumer profile: a description of the typical person or people who will use the product.

Market research

Market research techniques used by designers

- interviewing the client
- questionnaire (asking likely users of the product)
- analysis of existing products
- analysis of the styles of other products used by the target market
- collection of relevant measurements.

AQA Examiner's tip

- You should be able to name different methods of collecting information from the client.
- You should also be able to create a client profile from information gathered.

7.2 Product analysis

A successful designer will spend time looking at existing products. This is called product analysis. The designer will:

- investigate and study the functions and features of the products
- work out which parts of the products are most successful
- identify weaknesses in the designs
- work out how they can design new products that improve on existing ones.

Key features to analyse

Many designers look at products and analyse their features under selected headings.

In schools, some students use the acronym ACCESS FM, to remind themselves of eight key features to look at.

Some designers may use some slightly different headings such as:

Ease of use: video recorder v. Sky+ box

Ergonomics: Does the product fit people well? Is it comfortable to use?

Whatever headings are used, the most important thing is to use the findings of product analysis to influence and improve the design ideas produced.

Activities

1. Use the ACCESS FM headings to carry out a product analysis on a mobile phone.
2. Look at the photos opposite. On each row there is a photo of a traditional and a modern product with the same function. What weakness has the designer found in the original design?
3. In what ways has the new design improved on the original?

Summary

Analysing existing products helps a designer to invent an improved product.

A range of factors need to be considered when looking at existing products.

Objectives

Understand what product analysis is.

Know how it is used to improve new products.

ACCESS FM

A – Aesthetics

C – Cost

C – Client

E – Environment

S – Safety

S – Size

F – Function

M – Materials

AQA Examiner's tip

- Be able to analyse a product to describe its main design features.
- Know eight or more categories under which to analyse the product.

Key terms

Aesthetics: the look, and therefore appeal, of a product.

How product analysis has led to better design

A

Earlier design	Feature analysed	Improved design
Traditional lemon juice squeezer 	**Aesthetics** How we respond to the look of a product, its colour, texture, styling – old or modern, angular or curved. Is it appealing?	More attractive, Phillipe Starck designed squeezer 
Traditional coffee table that would appeal to older people 	**Client** Who is going to buy and use the product? What is the profile of the user?	Modern, sleek coffee table that would appeal to younger buyers 
These batteries are used once, and then disposed of 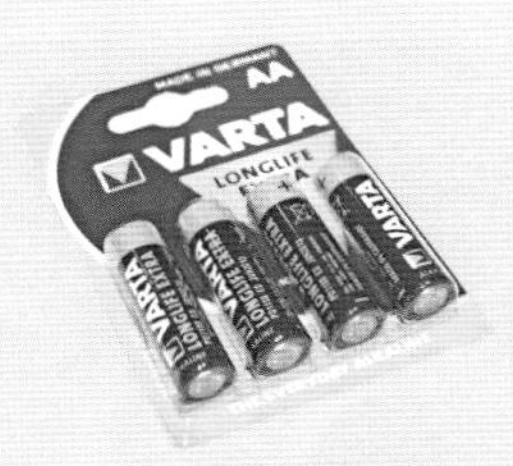	**Environment** Are the materials sustainable? What is the environmental impact of manufacture and use? Can it be recycled?	Wind-up radio needs no batteries 
Traditional children's playground equipment, with many potential hazards 	**Safety** Is the product safe to use? Is there a soft surface to land on? Are there safety straps? Do all the parts and components have smooth surfaces?	Modern, safer play equipment 
Traditional manual toothbrush 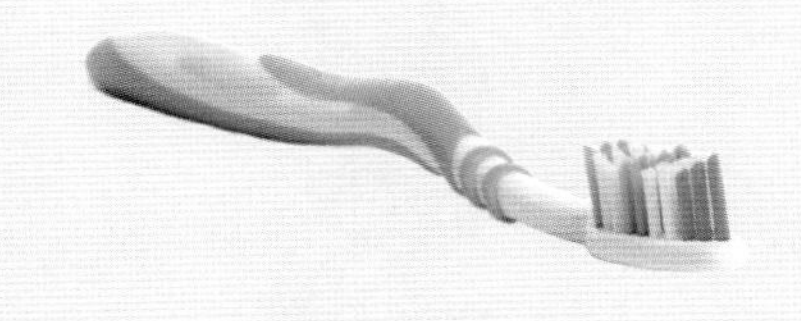	**Function** What is the product supposed to be used for? Does it work well?	Electronic toothbrush cleans more effectively 

7.3 The designer

The designer is a person, or team of people, who design and develop a product for a client.

A successful designer will analyse the design brief that the client gives them. They will carry out initial research to identify the following:

- Exactly what is the function of the product?
- Where is it going to be used?
- What sort of person is going to use the product? (Who is the target market?)
- What sorts of products are likely to appeal to this sort of user?
- What are the sizes of related items, for example, pens to be stored in a desk-top organiser?
- Are there existing products performing a similar function already being sold?
- What are the strengths and weaknesses of the existing products?

From the information gathered the designer will produce a more detailed brief or design specification. This will guide the designing that follows. The designer will aim to produce ideas that are creative. The designer will take into account all the **constraints** identified in research, and restrictions imposed by materials, manufacturing, legislation, etc.

Objectives

Understand the role of the designer.

Be aware of the constraints on a designer.

Be aware of the need to protect the ideas of a designer.

Key terms

Constraints: things that impose limits and boundaries to aspects of a product such as its size (size constraints) or the number of features (cost constraints).

Patent: this protects the features and processes that make things work. This lets inventors profit from their inventions.

Design teams

In larger companies there will be more than one designer. At a car manufacturer, different teams of designers will work on the body-shell, the suspension, brakes, interior, etc. (see Photo **A**).

A *Different parts of a car are designed by different design teams*

Being creative

Designers will look to produce original and appealing designs. These will lead to greater sales of the product. Creativity could be achieved by:

- adding to the functions of existing products
- making existing products more appealing by broadening the range of the product
- inventing a new product that has brand new functions
- designing products that have an unusual, attractive appearance.

Protecting creativity with patents

A designer creating an original innovative idea will not be happy if someone else copies their idea. If they can prove the new idea is theirs' they can apply for a **patent**. If this is granted, the designer can take legal action against anyone copying their idea. James Dyson was successful in taking Hoover to court for copying his bagless vacuum cleaner (Photo **B**).

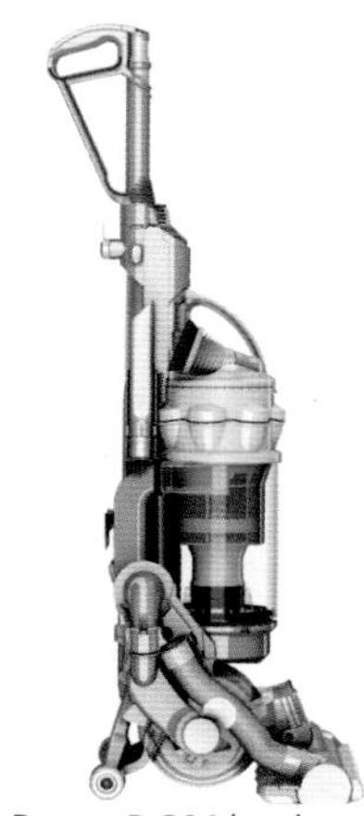

B *The Dyson DC01 bagless vacuum cleaner is protected by patents*

The table below shows how designers have focused on a particular feature in the design of chairs. As a result, each chair is very different.

AQA Examiner's tip

Be able to consider constraints when designing.

C *Design considerations*

Main design influence	Example	Main design influence	Example
Aesthetics Chair designed by Verner Panton, made from a single piece of plastic		**Anthropometric data** Child's high chair	
Cost Plastic chairs are moulded quickly and cheaply		**Environment** Recycled shopping trolley chair	
Ergonomics Häg Balans chair		**Function** Folding chair	
Materials Designer Frank Gehry's wiggle side chair from corrugated cardboard		**Use** Inflatable pool chair	

Activity

Study pictures of more chairs on the internet, or in catalogues. Find interesting designs and describe what you think has influenced the designer of each chair.

links

Look at the Nelson Thornes online resources to see more information on these chairs.

Summary

Know what a patent is, and why a designer might need to apply for one.

Know the factors a designer has to consider when designing.

7.4 The manufacturer

The manufacturer is the person or group of people who make a product or products. Examples of manufacturers include:

- a jewellery designer making individual pieces in a small workshop
- a small factory with injection moulding machines making a variety of plastic components in batches
- a large company assembling hundreds of cars every day on a production line in a large factory.

Objectives

Understand the role of the manufacturer.

links

For more see Quantity production on page 130.

Key terms

Quality control: a check made to ensure that a component meets the specification, for example correct size, shape and colour.

Quality assurance: a complete system of quality control checks and procedures throughout the manufacture of a product.

The manufacturer and designer working together

The manufacturer will have to work closely with the designer to ensure that the designer's ideas can actually be made. They will discuss:

- using standard components to reduce manufacturing costs
- reducing the number of components to reduce costs
- the skills of the available workforce
- the equipment needed to manufacture the product
- making sure the product complies with national and international standards
- the scale of production (how many are going to be made)
- how the product can be manufactured safely.

Product safety

Many products sold in the UK and in Europe have to meet strict standards. If a manufacturer believes their product meets all the European standards, they can put the CE symbol on the product (Picture **A**). This should make the product more appealing to consumers and increase sales throughout Europe.

The manufacturer can apply to have the product tested against the International standards for that product. In the UK, the British Standards Institute (BSI) is the body that carries out this testing. If the product passes this testing, the manufacturer is allowed to use the BSI kitemark (Picture **B**).

A *The CE mark*

The standards for children's toys

An example of standards that products must meet is the one for toys. BS EN 71 is the British and European standard. It is in eight parts, which apply to different sorts of toys, but include rules on:

- size and shape of parts
- flammability of parts
- chemicals used in the toy
- labelling
- maximum heights of play equipment.

B *The BSI kitemark*

Activity

Look at the diagram of the 240-volt electric hedge-trimmer (Photo **C**). Copy the sketch. Label the safety features that you believe have been included.

C *Electrical goods have to meet standards*

AQA Examiner's tip

Make sure you understand how the work of the BSI is essential both to the manufacturer and the consumer.

Quality control and quality assurance

When a product is made, each part has to be checked to ensure it is correct. It can be inspected, for example to make sure the colour is correct. It can be tested, for example to make sure that is the right size. The checks made are **quality control** checks. Many manufacturers will carry out these checks every hour, taking a sample from the production line. If the part does not meet standards, they will discard all of those parts made since the previous check.

A manufacturer needs to plan all the quality control checks throughout the manufacturing process. This includes checking the raw materials as they arrive at the factory right through to checking the finished product, its packaging, labelling and distribution. This total plan is **quality assurance**. The manufacturer is doing everything possible to assure the consumer that every single product will have the same high quality.

D *Moulded plastic chairs*

The cost of tooling up to manufacture

A manufacturer has to consider the costs of making new tools to make new products. An injection moulded plastic garden chair will need a very expensive mould (Photo **D**). Once made, the cost can be spread across all the chairs made. The more chairs made, the cheaper each one will be.

Another consideration for manufacturers is the cost of labour. Could manufacturing costs be reduced by using more machines and robots to make the product? Could the product be made in another country, where the labour costs are much cheaper? This is an example of global manufacturing.

Summary

A manufacturer makes products to sell.

Manufactured products should be tested to ensure consumer satisfaction and safety.

8 Presenting ideas

8.1 Drawing techniques

You need to be able to communicate your design ideas to other people. Two-dimensional (2D) and three-dimensional (3D) techniques can be used for sketching and drawing your ideas. You can use grids under your paper to help you with some of these methods. Drawings or sketches may be done freehand or using a ruler. The use of a ruler is not essential.

Objectives

Be aware of a range of 2D and 3D drawing techniques.

Be able to use some of these techniques to communicate ideas.

2D techniques

2D sketches and drawings show one face, view or elevation of an object, for example, the front view of a table or the face of a mobile phone (Diagram **A**). You would use 2D sketches and drawings to design parts to be cut on a laser cutter. Use a square grid under your paper to help with 2D sketches (Diagram **B**).

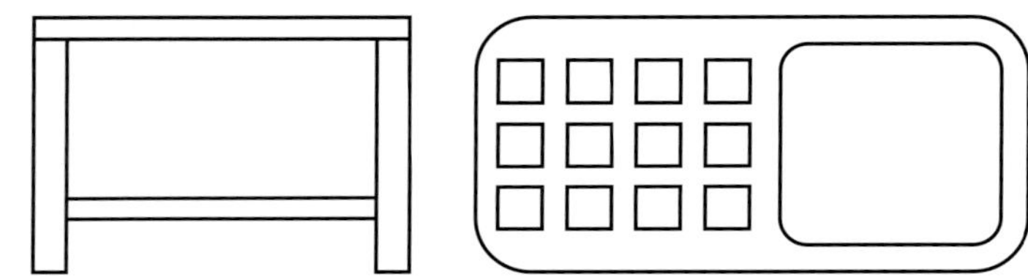

A *2D drawings of a coffee table and a mobile phone*

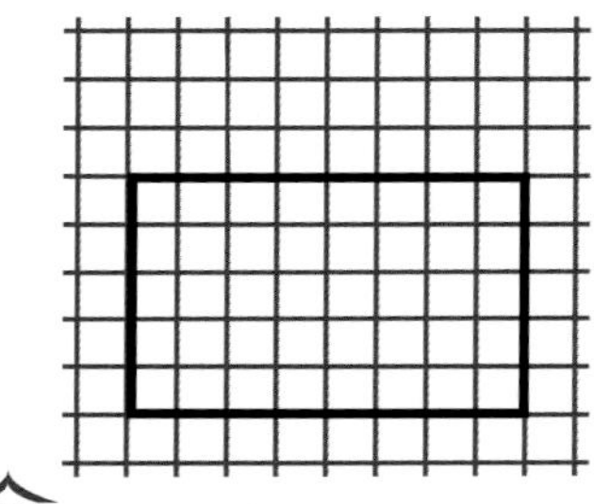

B *Square grid*

More advanced 2D techniques

If you wish to combine several 2D views of the same product on the same page, you need to lay them out as shown below (Diagram **C**). This is the most common layout of formal drawings. It is known as third angle orthographic projection. This formal orthographic drawing includes a cross section. A cross section is a useful technique that shows what a product would look like if you cut through it.

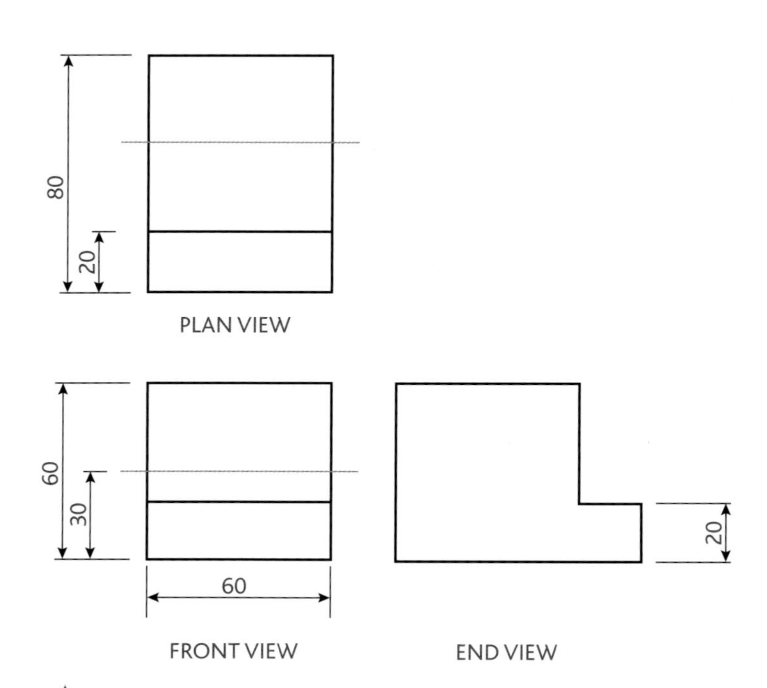

C *Formal orthographic drawing*

3D sketching

3D sketches have the advantages of showing three faces of an object at the same time. The disadvantage of this style of sketching is that some or all of the faces are distorted. Isometric sketching (Diagram **D**) is the most popular form of 3D sketching. Each face is distorted at an angle of 30°. Use an isometric grid under your paper to help you with isometric sketches (Diagram **E**).

links

Some CAD programs allow you to change 3D designs to formal orthographic drawings very quickly. See pages 96–97, Evaluating design ideas.

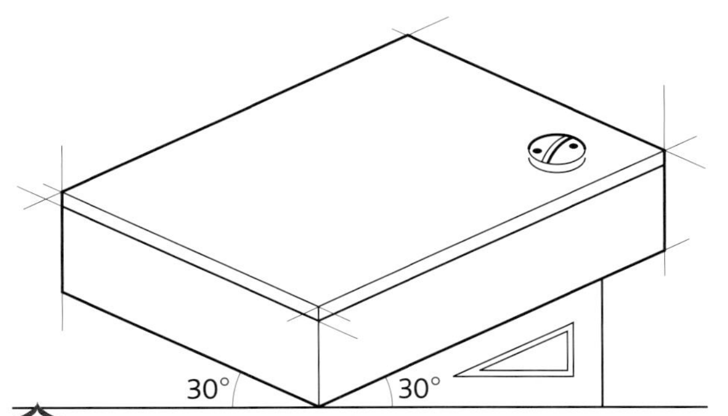

D *Isometric sketch. Isometric sketching is the most popular form of 3D sketching. Each face is distorted at an angle of 30°*

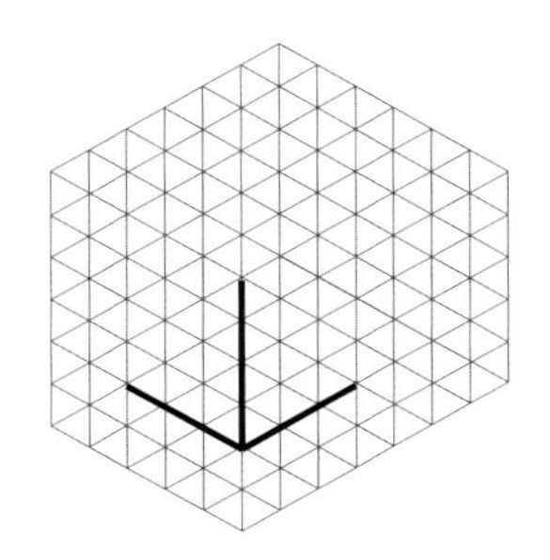

E *Isometric grid*

Other techniques that you could use include perspective and cabinet oblique projection. Perspective is more difficult to sketch quickly, but you may choose to use it for a presentation sketch of a final idea. It gives a more realistic view of the object.

Cabinet oblique is easier to draw than isometric, but does not show the object as well. The front face is not distorted, but the side and top are distorted.

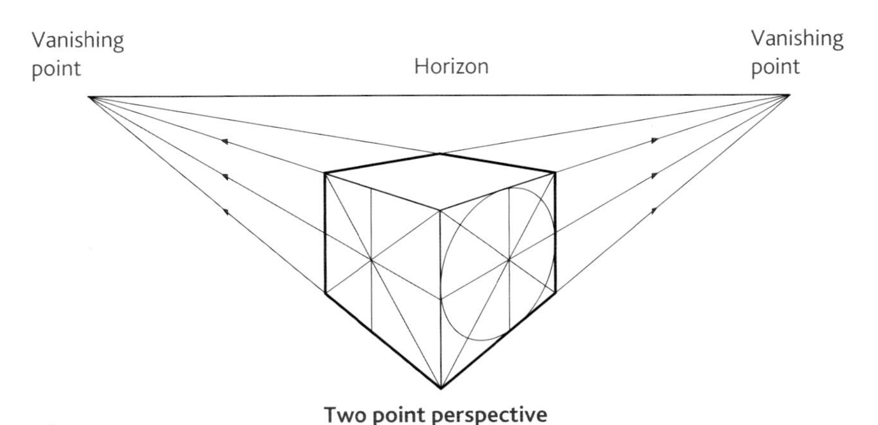

F *Perspective sketch. Perspective sketches are more difficult to draw, but are more realistic*

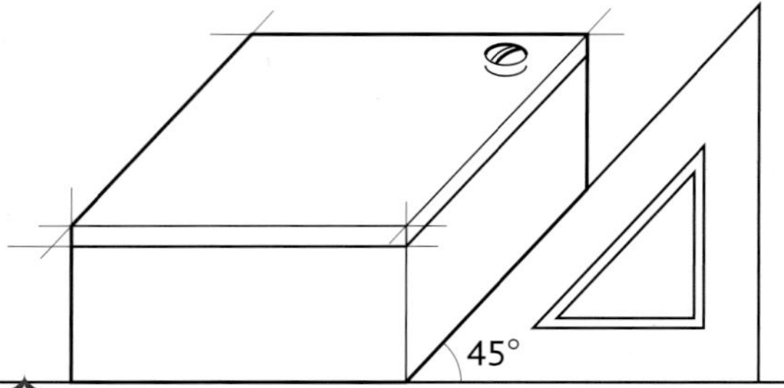

G *Cabinet oblique sketch. The front face is not distorted, but the top and side look distorted*

More advanced 3D techniques

To show clearly how parts of an object fit together, it is helpful to produce an **exploded view** of the product. All the parts are separated along 'lines of action'.

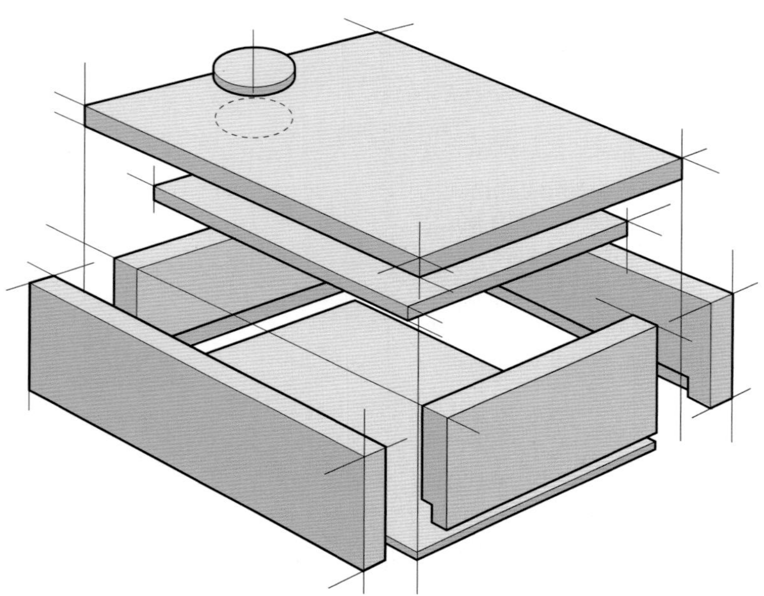

H *An exploded view of a box that has been rendered to show the materials*

Activities

1. Sketch a simple coffee table.
2. Sketch a cross section through the table.
3. Produce an exploded sketch of the coffee table.
4. Render the exploded view to show the materials.

Key terms

Exploded view: a 3D sketch or drawing where the parts of the product have been pulled apart and drawn in line with the parts to which they need to fit.

Rendering: adding colour to a sketch to help show the materials and textures of surfaces.

AQA Examiner's tip

Be able to communicate your ideas clearly using a range of 2D and 3D techniques. Practise so that you can produce sketches rapidly.

Rendering

Adding colour and shade to an object makes it look realistic. The colour can be applied to simulate different materials and finishes. This is called **rendering**. You can use coloured pencils, marker pens, pastels, airbrushing and paints.

Most CAD software allows you to render your designs. It can usually be done quickly, so a range of options can be explored rapidly.

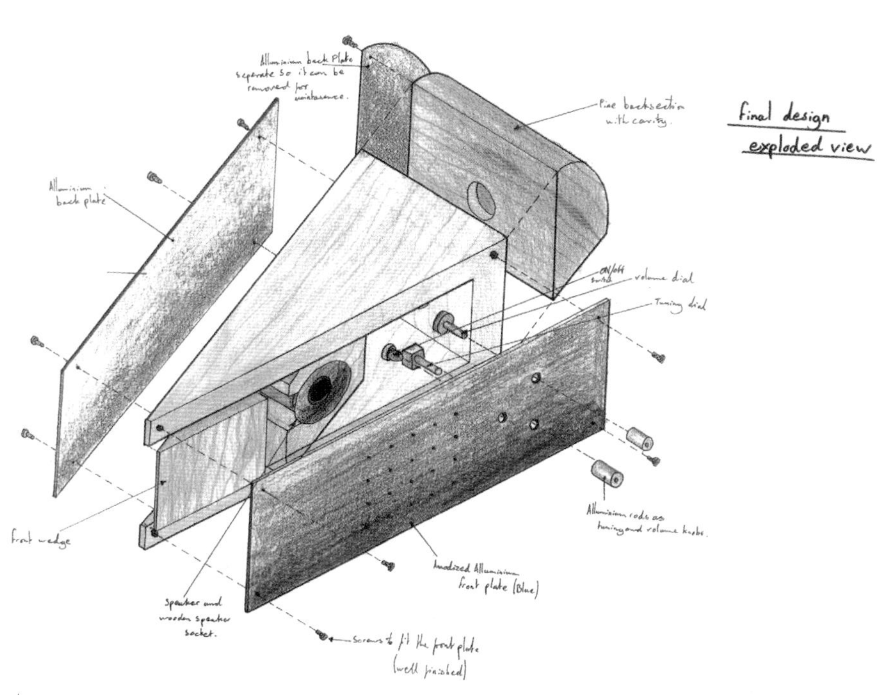

I *A full colour sketch*

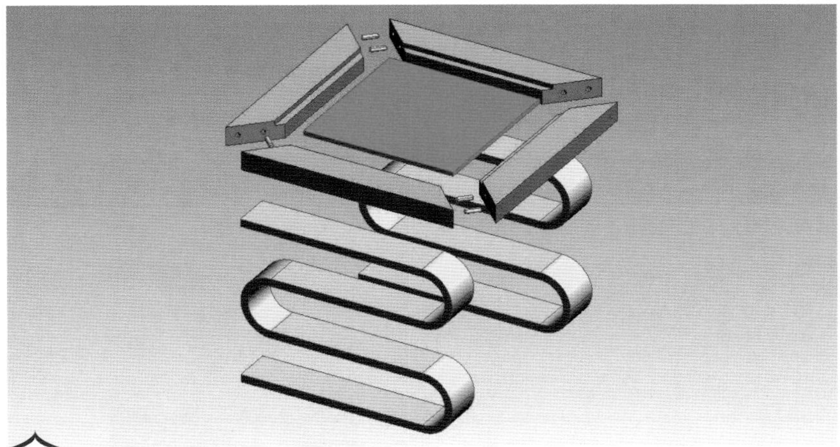

J *A rendered CAD design (Pro/DESKTOP)*

Summary

2D sketches are useful for showing detail.

3D sketches are useful for showing what the whole product will look like.

Exploded sketches show how a product can be assembled.

Rendering can enhance sketches.

Computer-aided design combines all the advantages of the techniques above.

8.2 Models and prototypes

Prototypes

Why do designers make prototype models?

James Dyson made 5,127 **prototypes** during the design of his cyclone vacuum cleaner. Some were working models, made to help improve the function of the cleaner. Some were non-working models, made to help the design of shapes of parts. All of his prototypes helped him to refine his revolutionary vacuum cleaner design. Dyson is a great believer in using rough physical models to **develop ideas** quickly. He also believes in 'stealing' technology from other applications. His inspiration for the dual cyclone vacuum cleaner came from a saw-mill. He crept in at night to study and sketch the 10-metre high tower that separated sawdust made by machines. He created his own version for his factory making ball barrows, and thought 'Why not scale it down for a vacuum cleaner?'

A *A prototype model used to develop a vacuum cleaner*

B *A model of a vacuum cleaner*

Objectives

- Be able to identify common modelling materials.
- Understand how models have been used to develop commercial products.
- Understand how to cut, shape and join common modelling materials.

Activity

1. Make a rapid model or models of ideas for a desk stand for a mobile phone. Use any available suitable materials.

links

View the range of models in Unit 2 Design and making practice.

Key terms

Prototype: a model of a product that is used to test a design before it goes into production.

Develop (ideas): simplify, refine, clarify, modify and decide constructional methods to improve your design.

What materials are suitable for prototype modelling?

C *Common modelling materials*

Material	Advantages	Image
Paper	Quick and easy to cut and fold Little structural rigidity	
Thin card	Quick and easy to cut and fold Better structural rigidity	

Corrugated card	Easy to source pre-used corrugated card More difficult to fold 
Styrofoam	Great for carving into solid block shapes This is one example of the uses of styrofoam 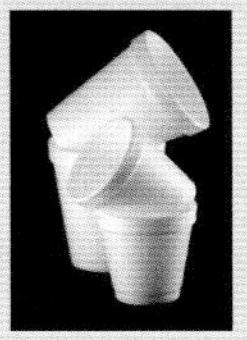
Foamboard	Can give clean, crisp models if carefully cut and shaped 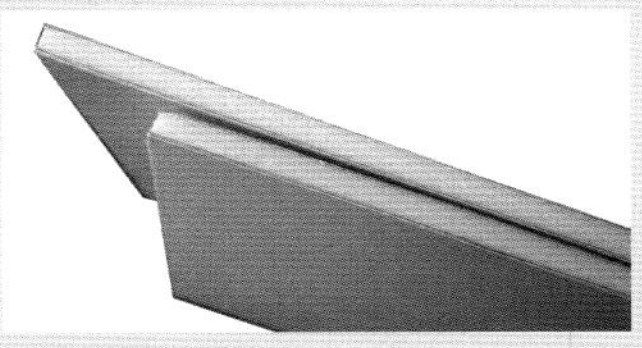
Balsa	Easy to cut and shape in a school workshop 
Wire	Can be bent easily into complex shapes 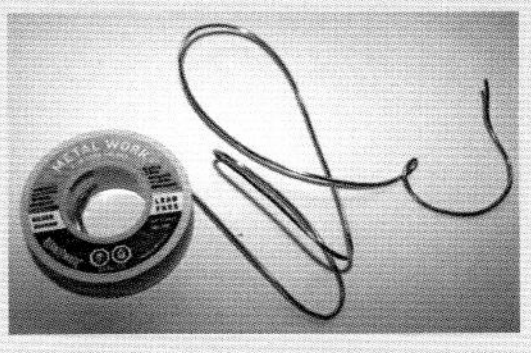
Straws	Great for representing tubes and for axle bearings Difficult to bend 
Pipe cleaners	Good for complex shapes 
Polymorph	Can be shaped more rapidly than styrofoam Can be reused 
Construction kits	Particularly good for modelling mechanical parts because of their accuracy 

How can you use models?

Producing a quick model is a good way to help you see your idea as a 3D shape. If you produce more than one model, you can improve your ideas. You can:

- work out how individual parts can fit together and how big they need to be
- refine the shapes of parts
- check that the design works
- show the model to your client and the consumer, to get their views.

Cutting and shaping materials

A cutting mat, craft knife and straight edge are good for use on many materials.

Other common workshop tools may be useful for specific materials, for example, pliers for bending wire, files and saws for cutting and shaping materials such as balsa and styrofoam.

Activity

2 Make further models to improve the design of the phone stand you made in Activity 1.

How can you make the stand so that the sound comes out of the phone's speaker more clearly?

links

Unit 2 Design and making practice, shows examples of how models have been used to develop ideas. See page 150.

You will find more information on polymorphs on page 24.

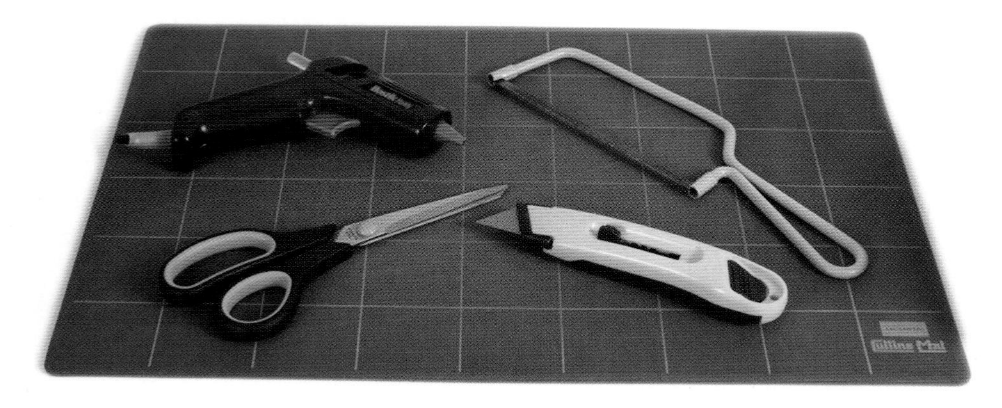

D *Common workshop tools used to cut, shape and join different modelling materials*

Joining materials

Hot glue is quick and easy to use to join a variety of modelling materials. Care needs to be taken to avoid burns.

James Dyson used gaffer or duct tape in lots of his working models as it is quick and strong. Staples and paper fasteners are also useful.

Remember

A laser cutter can be used to cut parts for a model quickly.

Examples of models

E *Coffee table model in card, dowels and clear self-adhesive tape*

F *Child's toy prototype from pipe cleaners and styrofoam*

G *A child's toy model in styrofoam*

H *A model from foamboard, overhead projector transparency and styrofoam*

I *Model of child's toy car in balsa*

Summary

Models and prototypes are used to show, test and help refine a design.

There are a variety of modelling materials available to a designer.

Modelling materials are quick and easy to cut and join together.

Modelling materials are relatively inexpensive.

AQA *Examiner's tip*

- Be able to describe the advantages and problems of using modelling to communicate ideas.
- In the controlled assessment task, you may need to produce many models to help you finalise your design.

8.3 Using ICT as a design tool

Computer-aided design (CAD) allows ideas to be presented on screen or on paper. Ideas can be modified rapidly, and shared through the internet instantly.

When the CAD software is linked to a CAM (computer-aided manufacture) machine, the design can be manufactured rapidly and accurately.

Objectives

- Understand how CAD can be used to present ideas.
- Understand how CAD can be used to develop ideas.

2D CAD

2D CAD software is excellent for showing one face of an object, for example the front. Shapes can be drawn with absolute accuracy. Drawings can be used to cut materials on a laser cutter or vinyl cutter.

Activities

1. Use CAD software to draw a mobile phone. Include all the detail, such as buttons.
2. Use CAD software to design docking stations with speakers for the phone you have drawn.

Software for 2D CAD

Techsoft 2D design is very popular in schools for 2D CAD. Some students have successfully used other software including Autocad, TurboCAD, Corel Draw, and even Paint and Publisher.

Creating formal orthographic drawings

You will usually need at least three different views of the product you design. The top view should be above the front view, and the side view should be beside the front view. This is an orthographic drawing.

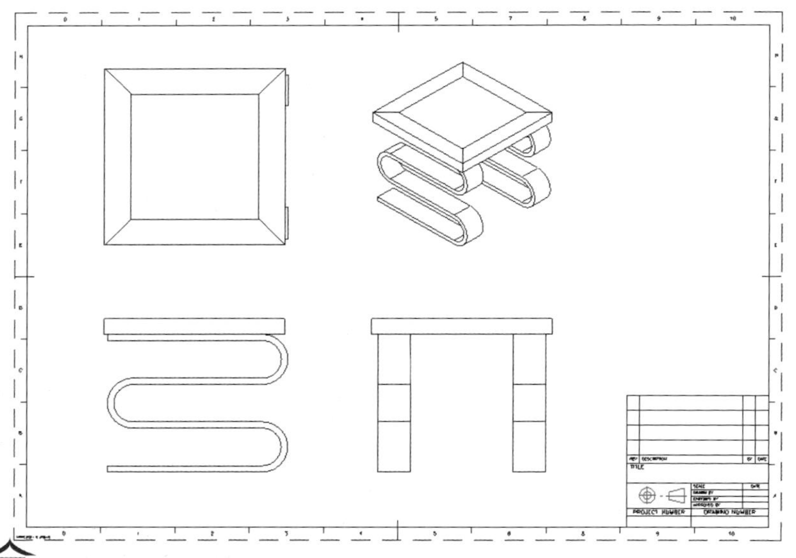

A *A formal orthographic drawing*

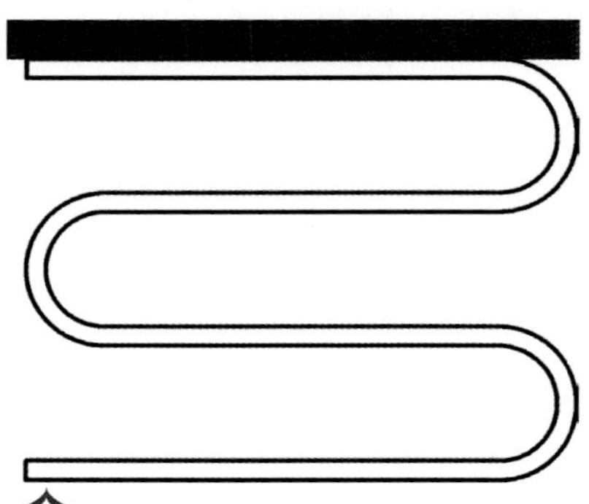

B *Side view of a coffee table drawn using Techsoft 2D Design V2*

3D CAD

3D design software has the advantage of allowing a designed object to be rotated so that it can be seen from any angle. Designs can be sent to a 3D printer (rapid prototyping machine) to produce a solid model.

More advanced 3D CAD techniques

The example above was drawn as a single part. The diagram below shows the same object drawn as separate parts. These have been **assembled**. Assembled drawings can be pulled apart (exploded) to show construction methods.

As your CAD skills improve, you can create a range of ideas on screen. You can also use CAD to develop, refine and improve your ideas. Most 3D CAD software packages allow you to create formal orthographic drawings in seconds from your design. Some packages allow you animate objects and carry out stress analysis on parts of the design.

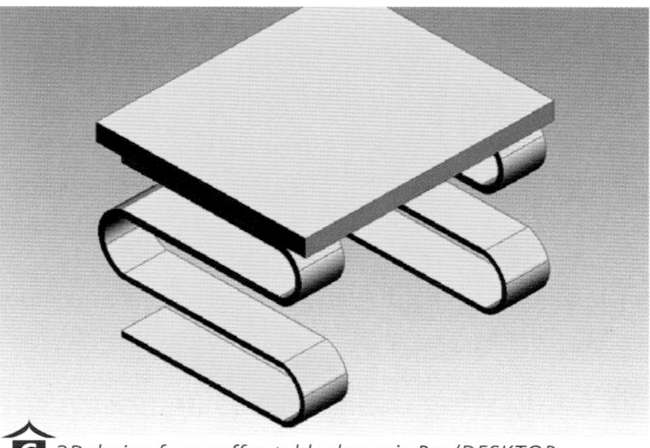

C *3D design for a coffee table drawn in Pro/DESKTOP*

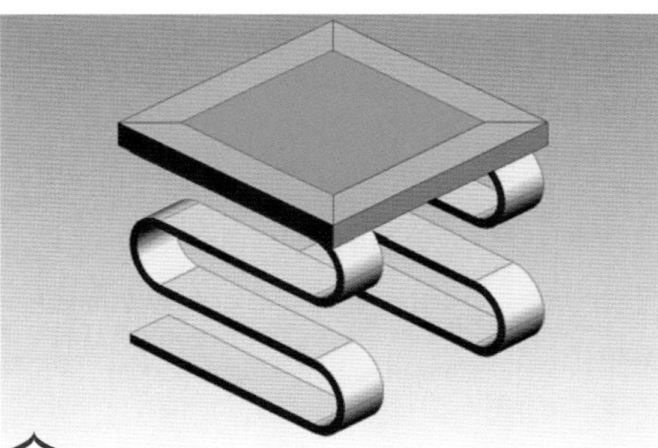

D *This table has been drawn as separate parts, which have then been put together*

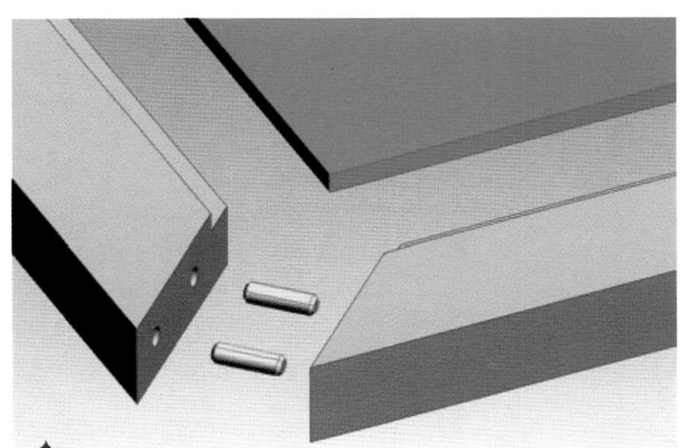

E *An exploded view of the corner of the table. Hidden dowels are shown*

Summary

CAD is computer-aided design (either 2D or 3D).

CAD is used to present and develop design ideas.

Software for 3D CAD

Google SketchUp, Pro/DESKTOP, ProEngineer and Solid Works are all popular 3D CAD software used in schools.

links

For more examples of the use of CAD to develop ideas, see Unit 2 Design and making practice.
For videos of 3D CAD advanced techniques, see the *Kerboodle!* online resources.

Key terms

Assemble (in CAD): place together separate parts of an object drawn in computer-aided design.

AQA Examiner's tip

Be able to explain the advantages of using computer-aided design software in the design process.

8.4 Evaluating design ideas

A designer must study the needs of the client and consumer before generating design ideas. They will have listed these needs as a design specification.

A designer will then need to **evaluate** their design ideas, to check how well each idea fits the design specification.

It is also useful to get the client's reaction to design ideas and reactions from likely users of the new product. This is important because:

- it is possible that the designer may misunderstand the needs of the client
- it is possible that the client and/or consumer do not find the ideas appealing.

Objectives

Understand why design ideas need to be evaluated.

Be aware of what methods can be used to evaluate ideas.

Key terms

Evaluate: make judgements about how successful something is.

Evaluating design ideas against a specification

A designer will possibly evaluate their design ideas without making notes. When you work on your controlled assessment task, you will need to show evidence of how you have evaluated your design ideas against the design specification. You could:

- make a checklist from your specification and tick off each point that the design meets
- add notes to your design sketches commenting on how well each design meets particular parts of the specification
- analyse why your design does not meet all the needs of the specification
- explain why you have had to make compromises in your design.

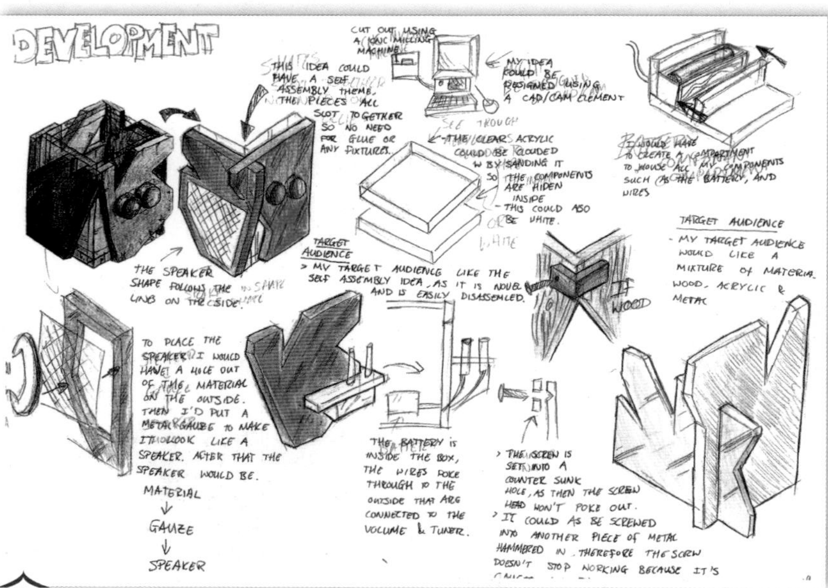

A *Part of design ideas sketches showing how there has been on-going evaluation against specification points*

links

Further examples of evaluation of design ideas can be found in Unit 2 Design and making practice.

Activities

1. Carry out a survey to find out what features and functions teenagers look for in a phone.
2. Write the most popular requests as a list or specification.
3. Research using the internet to find five phones that best match this specification.

Presenting ideas to a client/consumer for evaluation

In order to get consumer reaction to ideas, ideas need to be presented in a way that the consumer can understand easily. This could be by using:

- clear 3D and 2D presentation sketches and notes
- a rendered 3D computer aided design (CAD)
- 3D physical models made by the designer
- 3D models made by a CAD/CAM machine
- a combination of any of the above methods.

Remember

When you are asking consumers to evaluate your own design ideas, you will need to record their reactions in your design folder. You may need to modify your design ideas because of the comments they make.

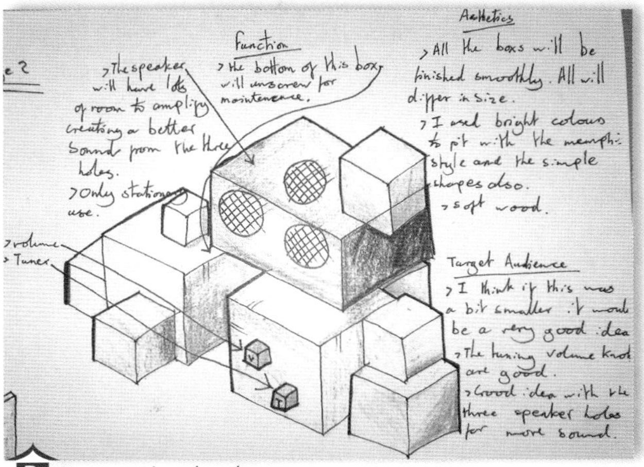

B *Presentation sketches*

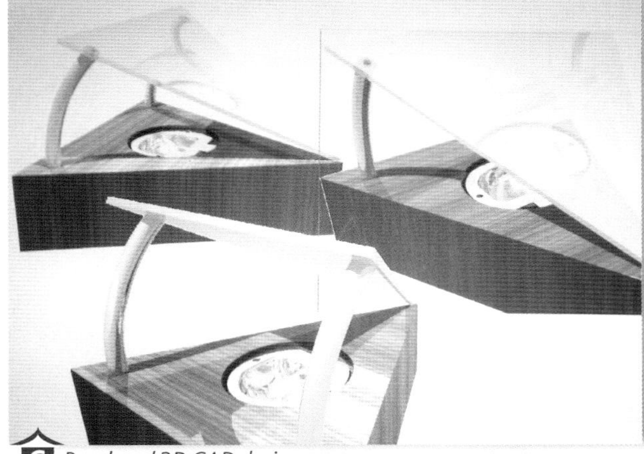
C *Rendered 3D CAD design*

D *3D model*

E *Using a laser cutter (CAD/CAM)*

Summary

Design ideas need to be evaluated by the designer, and by the client or likely users of the product.

A range of techniques can be used to present ideas to a client or users.

AQA Examiner's tip

When evaluating design ideas in the design question, remember to refer back to the specification given.

8.5 Planning for manufacture

Formal (engineering) drawings

When you have finalised your design for a product, you need to be able to plan how to manufacture it.

In Unit 2 Design and making practice, you will need to produce accurate sketches or drawings of the final design. Computer-aided design may be useful for this.

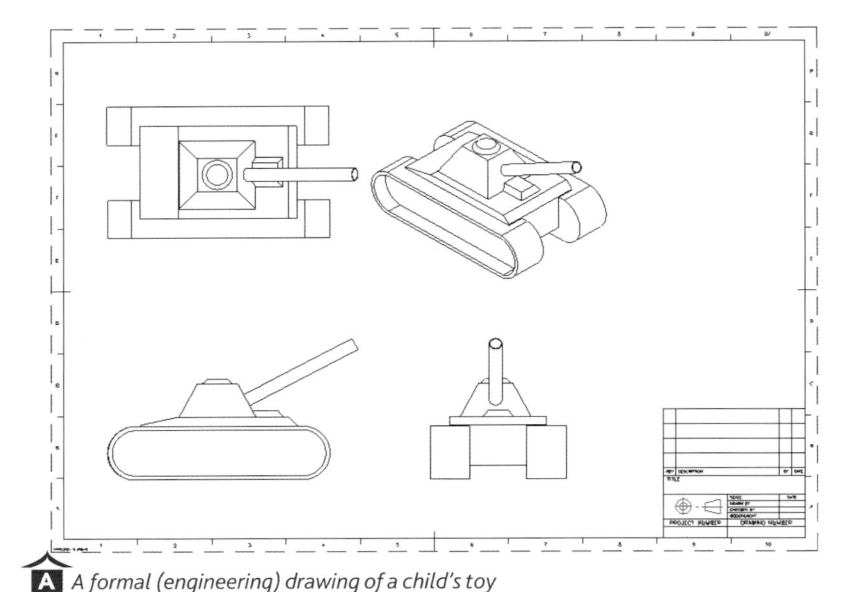

A *A formal (engineering) drawing of a child's toy*

Plan for manufacture (plan of making)

In both Unit 2 Design and making practice, and in your written examination, you have to produce a plan of manufacture. This means sorting the main sequence of making activities into a sensible order. In the examination, you will need to be able to sketch the main operations of making.

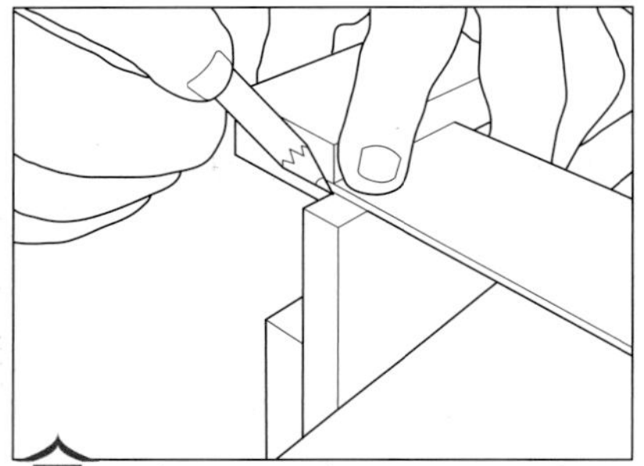

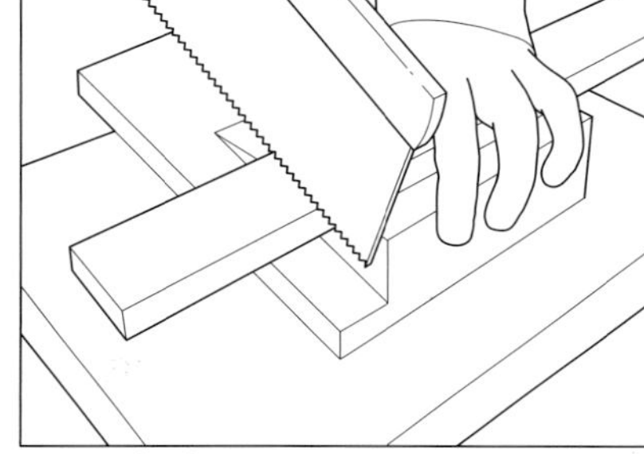

B *Marking out and cutting stages for making a wooden box*

Objectives

Be able to organise the main activities of manufacturing a product into the correct sequence.

Remember

If you have used Pro/DESKTOP to produce 3D designs, formal engineering drawings can be created in seconds from the 3D design.

Key terms

Flowchart: a list of processes organised into the best order. Decisions need to be made at stages in the flowchart. Arrows show the route through.

links

Try out the exercise on the Kerboodle website to organise making tasks into the right order.

Including quality control

You will also need to include quality control checks. These are the times during the making when you need to check your work is correct. An example of a quality control check could be making sure two pieces of wood fit together without gaps when making a joint.

Using a flowchart for a plan of making

A **flowchart** is an excellent way of laying out a plan of making. Each step in the sequence is put in a separate rectangular box. Each quality check is put in a diamond shaped box. The boxes are then linked with arrows.

Producing a cutting list

Part of a plan of making is a list of the materials you will need for your project. You will also need to specify what finishes will need to be applied to the materials.

C *A cutting list for a simple coffee table*

Part	No off	Length	Width	Thickness	Material	Finish
top	1	600	450	15	MDF	Emulsion paint applied with a roller
legs	4	420	40	40	Pine	3 coats polyurethane varnish
long rail	1	500	40	20	Pine	3 coats polyurethane varnish
short rails	2	380	40	20	Pine	3 coats polyurethane varnish
shelf	1	520	360	15	MDF	Emulsion paint applied with a roller

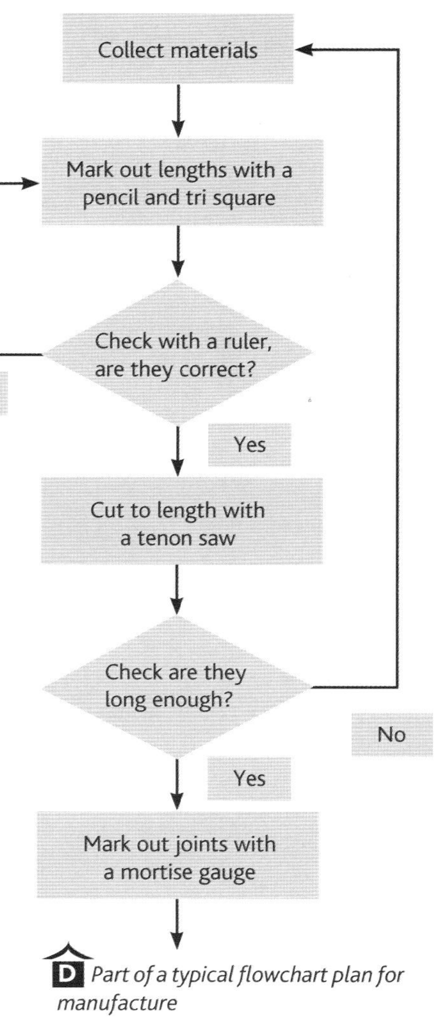

D *Part of a typical flowchart plan for manufacture*

Activity

Look at the steel tea-light holder. Write a list of the steps to make this product in the workshop. Rewrite the list as a flowchart.

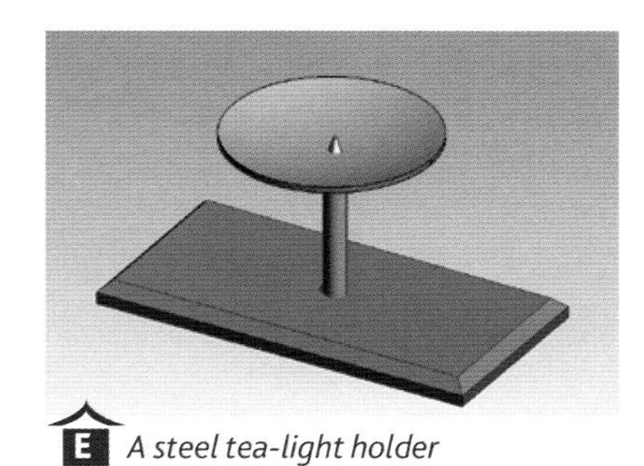

E *A steel tea-light holder*

AQA Examiner's tip

Be able to sketch the main steps involved in making a product, for example, marking out, cutting, joining, forming and finishing, or CAD/CAM equivalents.

Summary

Accurate formal (engineering) drawings or sketches are needed as part of a plan of making.

A flowchart is an excellent way to list the tasks in a plan of making.

A cutting list should include details of sizes, materials and finishes for all the parts of a product.

AQA Examination-style questions

Test your understanding of design and market influence by answering the following exam-style questions.

Inspiration and innovation

1 Study the two products shown below.

A **B**

a) Which one has developed from 'market pull' and which one has been developed as a result of 'technology push'? *(2 marks)*

b) Explain the reasons for your choice. *(8 marks)*

2 A manufacturer of jewellery has asked you to design a range of jewellery based on the theme of the natural world. Produce a range of ideas for a brooch that is based on an insect. *(10 marks)*

3 Produce a design for a chair in the style of Charles Rennie Mackintosh. *(6 marks)*

4 Study the Xbox controller shown below.

C

a) Identify and describe **five** ergonomic features of the Xbox controller. *(10 marks)*

b) Explain how anthropometric data has been used in the designing of the Xbox controller. *(4 marks)*

Design influences

5 a) Produce a labelled sketch of a product that has been designed for an inclusive target market. *(6 marks)*

b) Explain the reasons for your choice. *(4 marks)*

6 a) Produce a labelled sketch of a product that has been designed for an exclusive target market. *(6 marks)*

b) Explain the reasons for your choice. *(4 marks)*

Sustainability

7 Name and describe each of the 6 Rs. *(12 marks)*

8 Describe how aluminium is produced from its raw material. *(6 marks)*

9 Discuss the sustainability issues associated with producing and using plastic carrier bags. *(10 marks)*

10 Explain why it is important for designers to consider maintenance when designing expensive products, such as motor vehicles. *(6 marks)*

Client, designer and manufacturer

11 Give **three** methods of gaining information from a target market. *(3 marks)*
AQA specimen question

12 Describe the role of the designer in the manufacture of a product. *(3 marks)*

13 Explain how the scale of production affects the unit cost of a product. *(8 marks)*
AQA specimen question

Presenting ideas

14 Give **three** graphical methods of producing ideas. *(3 marks)*

15 Explain the advantages and disadvantages of using CAD (computer-aided design) rather than traditional methods of drawing. *(8 marks)*

16 Explain why designers often produce a model of their design as part of the design process. *(6 marks)*

17 Produce a flow diagram to clearly show the stages of manufacture involved in making the joint shown below. The joint is made from pine wood. *(12 marks)*

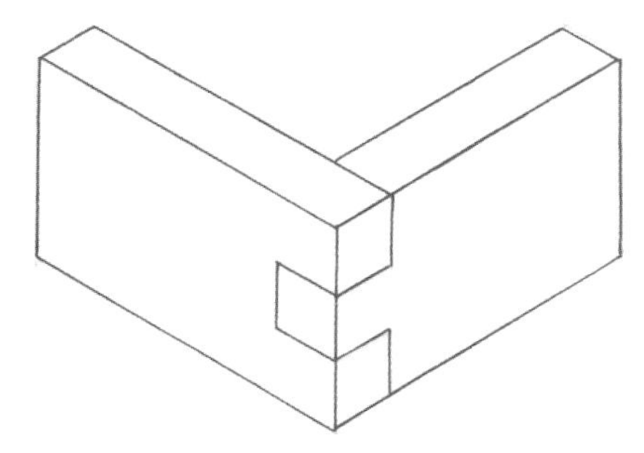

D

UNIT ONE Written paper

GCSE

Processes and manufacture

In these chapters you will learn about:

- workshop safety
- control of substances hazardout to health (COSHH)
- hand tools
- power tools
- marking out
- joining wood
- joining metal
- casting
- forming wood
- deforming metal
- moulding plastics
- computer-aided manufacture (CAM)
- quantity production.

In the first part of this book you studied materials and the reasons why designers and manufacturers need a good understanding of the different types of materials. In these chapters you will see the importance of using the correct manufacturing technique to achieve products that are cost effective and of the highest standards possible. As you develop your own design ideas it can be really useful if you begin to think about how your idea might be made and the techniques and processes that you feel comfortable using and that are available. By understanding a variety of manufacturing processes you will be able to select the most suitable materials and tools to make your design ideas efficiently.

- You will learn about a whole range of different traditional tools and processes.
- You will learn about different power tools and the advantage they offer in manufacturing.
- You will find out about terminology relating to specialist equipment and will learn how to explain how materials change as they are processed.
- You will learn that even when working with different materials many of the techniques you use are very similar, that is, marking out and measuring.
- You will learn about workshop safety and how potential hazards can be identified and managed.
- You will find out about methods of manufacture controlled by computers that can improve accuracy and consistency in your making.

When you want to make your design you will need to consider constraints you might have with availability of materials and equipment. If you understand the basic processes of marking out, cutting/shaping, joining and forming then you can plan and select the best tools and processes to carry out your making. Many of the more traditional hand manufacturing techniques require practice to achieve a high standard of accuracy but if you decide to design and utilise appropriate jigs and formers you can achieve a high standard of accuracy in your making but with a lower skill level.

You may decide that you have an opportunity to make your design idea in larger quantities or may have to repeat a process (for example four sides of a box) and therefore you will need to study the section about quantity of manufacture and the use of jig, templates and formers. When you work on your controlled assessment task, the design of the jigs, etc. can sometimes be demanding and their inclusion into your work may increase the level of originality and complexity in your work and help you to access higher grades.

These chapters will clearly show you how to manufacture in a safe manner and that safety must be considered at all times. Safety in the school workshop is dependent not just on wearing protective equipment and behaving in a sensible manner but also involves procedures for storing, using and dealing with potentially dangerous materials.

And finally, don't forget to go online where you will learn more about the topics covered here and have the chance to test your knowledge with the exciting interactive tests.

9 Health and safety

9.1 Workshop safety

Most accidents that happen in a school workshop are caused by human carelessness. You must always concentrate on working safely, both for your safety and that of others using the workshop. One brief lapse of concentration could lead to an accident that changes your life, or that of a friend, for ever. It is your responsibility to behave in a mature and responsible manner, maintain a safe environment and use safe working practices.

Sensible behaviour is just one factor that improves safety in the workshop. Using **personal protective equipment (PPE)** reduces the risk of damage to your sight, hearing, breathing and skin. Each piece of equipment is designed to prevent certain hazards damaging you and to reduce the risk of personal injury.

Objectives

- Understand the need for health and safety.
- Be able to relate practical activities to the use of personal protective equipment.

Key terms

Personal protective equipment (PPE): for example goggles to protect the eyes.

A *Standard PPE used to reduce personal hazards*

Operation	PPE	Hazard	Safety symbol
Drilling, sanding, welding	Goggles, welding visor	Dust, swarf or sparks might get into eyes	Eye protection must be worn
General workshop activities	Apron	Clothing could be caught by tools or in machinery Dust and chemicals can get spilt onto clothing	Protective clothing must be worn in this area
Handling hot/ sharp materials	Heat-proof gloves, leather apron (plus leather leggings and full face mask if aluminium casting)	Burning hands/fingers when working with hot materials Cut hands when carrying sheet material such as steel or glass	
Using machinery	Ear defenders	Damaged hearing from repetitive or continuous loud noise	
Sanding, applying a finish, using adhesives	Face mask, latex gloves	Lung damage from inhaled dust or fumes	Wear face mask
Carrying or installing equipment	Stout shoes with toe protection	Damaged or crushed toes and feet caused by falling materials or equipment	Foot protection must be worn

Creating a safe working environment also involves being aware of the potential hazards (heat, chemical, electrical and dust) and managing and reducing the risk. The following points help create a safe environment for all users in a workshop.

- Carry tools and materials in a safe manner with sharp points and edges pointing away from people.
- Keep areas between benches and machinery clear to avoid tripping.
- Only use tools and machines that you have been shown how to use. Always use the correct PPE.
- Clean all equipment thoroughly after use.
- Always know where the emergency stop buttons are in the workshop (Pictures **B** and **C**).
- Tools should always be sharp and set correctly. When using sharp tools point them away from your body and keep both hands behind the cutting edge.
- Machinery should be regularly maintained to ensure guards are working, there is no electrical hazard and all parts of the machine function properly.
- Workshops should be well-ventilated with good dust extraction and good lighting levels.
- Only use chemicals after reading all instructions for use. Ask your teacher about disposal of chemical waste.

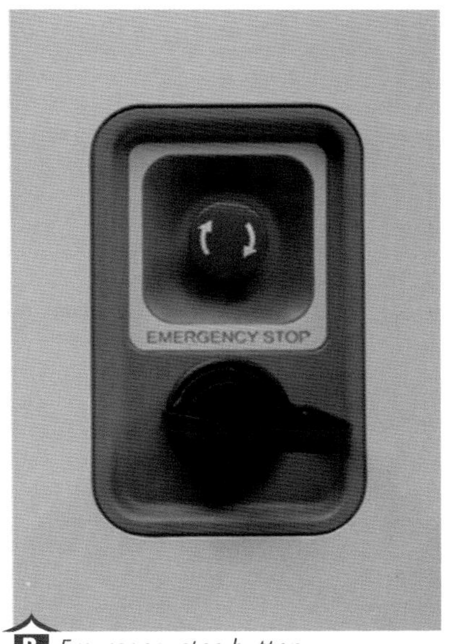

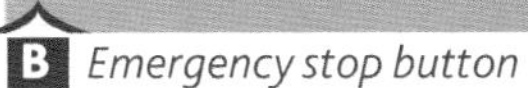
B *Emergency stop button*

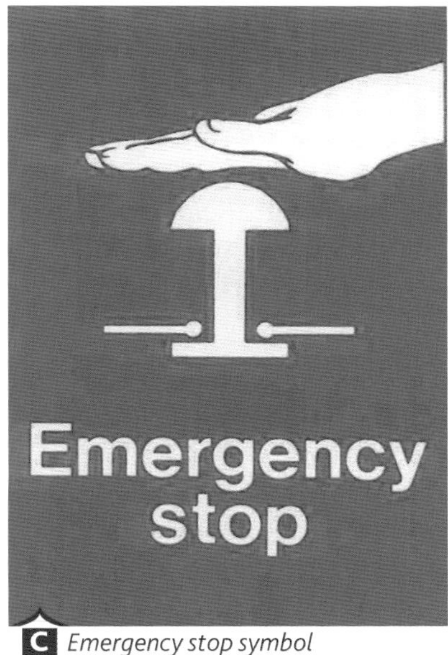

C *Emergency stop symbol*

Activities

1. Draw your workshop and identify any potential hazards to you and other users.
2. Identify five risks in your school workshop. Explain how the level of the risks have been managed and reduced by presenting your findings in a table format.

Summary

Every user in a school workshop should act and work in a way that maintains a safe and healthy environment.

Practical activities can present several risks to the user, which can be reduced by the use of PPE.

Accidents in a school workshop can be prevented by identifying hazards and reducing the risk to all users.

AQA *Examiner's tip*

- You need to explain the importance of PPE.
- You should be able to describe what PPE is relevant to a variety of activities.

9.2 COSHH

In a school workshop there are a number of hazards that are potential risks to all users. **COSHH** is a set of regulations (identified by COSHH symbol, picture **A**) that ensure hazards are controlled so as not to affect someone's health. Hazardous substances that you are likely to find in a school workshop are classified as substances you might use in work activities (for example, adhesives, paints, cleaning agents); and substances generated during work activities (for example, fumes from soldering and welding, and dust from sanding machines).

If these hazards are not controlled properly then after prolonged contact they are likely to have the following effects:

- skin irritation or dermatitis as a result of skin contact
- asthma or lung damage as a result of developing an allergy to certain substances
- losing consciousness as a result of being overcome by toxic fumes from solvents etc.
- cancer, which may appear long after the exposure to a certain toxic chemical that caused it.

Reducing risks with control measures

All control measures within COSHH regulations aim to prevent or adequately control exposure by:

- changing the process or activity so that the hazardous substance is not needed or generated, for example, use flux-free solder to prevent fumes being generated when connecting electronic components
- replacing high-risk substances with safer alternatives, for example, use water-based rather than solvent-based adhesives
- using substances in a safer form, for example, gel-based paints instead of paint in liquid form
- providing adequate ventilation and extraction of dust and fumes
- using a secure location for storage of harmful substances, for example, lockable cupboards to hold solvents and applied finishes
- providing good personal protective equipment (PPE) for eye/breathing protection, head protection, hand/arm protection and body protection.

Activity

1 Identify a number of procedures or machines that are used in your school workshop that help to control hazardous substances.

Hazard symbols

There is a wide range of hazard symbols, which are used quickly and easily to identify hazardous situations or materials. Symbols are often printed on individual containers (metal cans or lorries) that hold hazardous materials and in locations where hazardous substances are stored.

Objectives

Understand the need for health and safety.

Have a working knowledge of how to handle, use and store a variety of hazardous substances.

Key terms

COSHH: Control of Substances Hazardous to Health.

A *COSHH symbol*

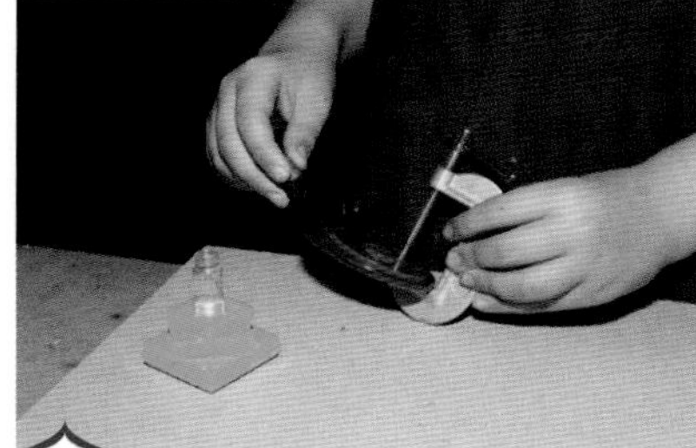

B *Cementing acrylic together with Tensol cement*

Remember

- Always read instructions on the containers of chemical substances like glue and solvents.
- Follow the instructions with great care, paying special attention to any warnings.
- Clean away and dispose of any waste quickly and safely, following instructions.
- Wash hands thoroughly after any work.

C *Hazard symbols*

Symbol	Meaning	Hazard	Control measure
	Flammable	The substance will have a flashpoint equal to or greater than 21°C and therefore will catch fire easily.	Use with great care in a well-ventilated area and keep away from naked flames.
	Toxic	The substance is poisonous and if it is inhaled, swallowed or if it penetrates the skin, may involve serious, acute or chronic health risks and even death.	Avoid use if possible but only use with PPE and only use in very small quantities under close supervision. Dispose of waste with great care.
Danger harmful	Harmful	A substance that is less dangerous than toxic substances but can cause inflammation if it is inhaled or ingested or if it penetrates the skin.	Avoid use if possible but only use with PPE and only use in very small quantities under close supervision. Dispose of waste with great care.
	Corrosive	A substance that is immediately dangerous to the tissues they contact and will attack and destroy living tissue, including the skin and eyes.	Avoid using large quantities. Only use with correct PPE and use in a safety container/tray to avoid spills. Dispose of waste with great care, following instructions on the product.
IRRITANT 3	Irritant	The substance is not corrosive but may cause reddening, irritation or blistering of the skin and eyes and breathing problems.	Use the correct PPE and use in a safety container/tray to avoid spills.

Activity

2 List three substances you might find in a school workshop that pose a risk to users. Produce a list of instructions to show how you should manage and reduce the risks you have identified.

Summary

There are hazardous substances present in a workshop that should be identified by specialist symbols.

Hazardous substances present risks to people's health, and consideration must be given to the use of alternatives if possible.

Handling, using and storing a variety of hazardous substances requires the use of control measures to reduce the risk of exposure to users.

AQA Examiner's tip

- You should be able to identify hazardous materials used in a school workshop.
- You should also be able to describe the precautions needed when using and storing common hazardous substances in the workshop.

10 Tools and equipment

10.1 Hand tools

Many of the hand tools you use in the school workshop have evolved over hundreds of years. Craftspeople through the centuries have developed the ability to use these simple tools to create both beautiful furniture and buildings. Only some hand tools are covered on these pages but they should help you understand the importance of using common tools safely and correctly.

Sawing

There are a number of saws available that are used for **wasting** and shaping materials. The teeth of the saw are slightly bent outwards, which provides the necessary clearance to prevent jamming but consequently the width of the cut is wider than the thickness of the blade.

A *Types of saw*

Tool	Material	Process
Tenon saw	Wood	The blade is stiffened to make straight cuts. It is used to cut pieces of wood to the correct length and wasting unwanted material.
Coping saw	Wood and plastics	The thin blades allow you to make curved cuts. The blade is held in tension by the spring steel frame with the teeth pointing backwards towards the handle. The angle of the blade can also be adjusted. Plastics should be cut slowly to reduce the heat caused by friction as this can melt the plastic and trap the blade. This problem can be reduced by sticking a piece of tape on the surface before you start cutting.
Hacksaw	Metal and plastics	Hacksaws have finer teeth and are mainly used for cutting metals. The blade of the hacksaw has the teeth pointing towards the front and is tensioned by the screw at the front of the hacksaw.
Junior hacksaw	Metal and plastics	This is a smaller version of the hacksaw with both a smaller blade and lighter metal frame. It is used to cut smaller/thinner pieces of metal and plastic and the blade is held in tension by the metal frame.

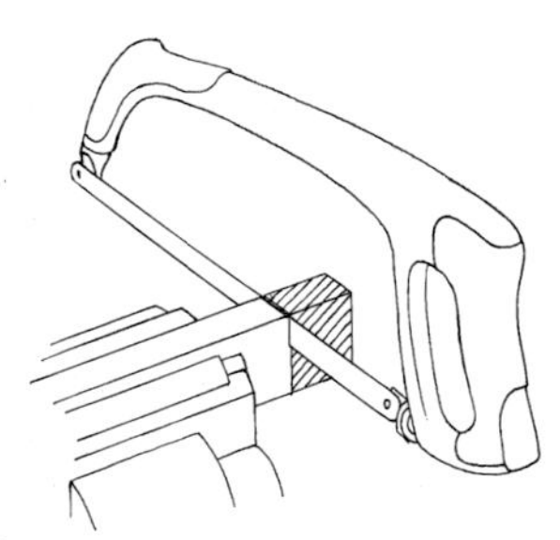

B *Correctly removing unwanted material*

Objectives

Have a working knowledge of a variety of hand tools and the particular processes they relate to.

Remember

Practice with hand tools will help you achieve a high standard in your making.

Key terms

Wasting: the mechanical removal of unwanted material by use of tools or machinery that use a cutting action.

Remember

- All marking out lines should be accurate and visible.
- Always use a vice or clamp to hold material securely before beginning to cut or chisel.
- To improve accuracy, always make cuts in the unwanted material and next to the marking out line.

AQA Examiner's tip

- You will need to know about a number of hand tools and understand how to select the correct tool for particular uses.
- You should be able to describe the correct method for using hand tools and any relevant safety procedures that are necessary.

Shaping

Materials can be shaped using a variety of hand tools that use some form of cutting action to remove material. Keeping tools sharp is important, as well as using them correctly, if a quality finish is to be achieved on the material. Wood and metal chisels use a basic wedge cutting action while planes use a similar action but the blade is held at a particular angle. Files use rows of teeth to remove small particles of material called filings.

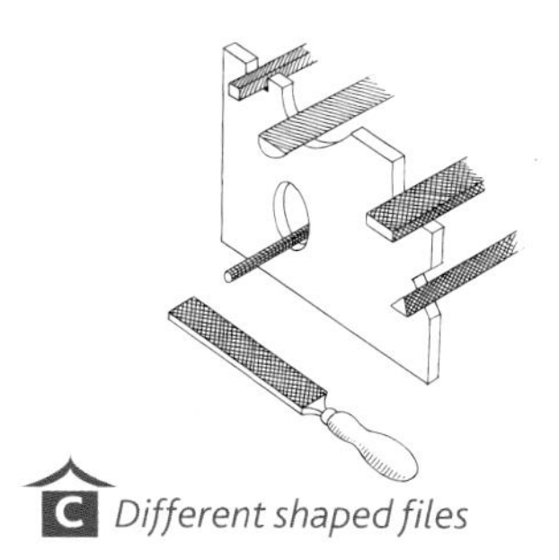

C *Different shaped files*

D *Tools for shaping materials*

Tool	Material	Process
Wood chisel	Wood	Chisels can be used for removing waste wood as well as shaping. Generally a firmer chisel is for general purposes, a bevel-edged chisel for awkward corners and a mortise chisel for chopping out mortises. A wooden mallet can assist with chiselling but should be used with gentle blows.
Cold chisel	Metal	The cutting edge of a cold chisel is much harder than a wood chisel. It is important that you always wear safety goggles when using a cold chisel.
File	Metal, plastic	Files are made from a very hard type of steel and are used for shaping metals and some plastics. They come in various sizes and shapes (diagram C) and are used in two basic actions – cross-filing and draw filing. Cross-filing is used across the material and is used to reduce material to a line. Draw filing is used to produce a smooth, shiny finish.
Plane	Wood	Planes are used mainly with wood to remove shavings and reduce the size of the material to a marked out line or marking gauge line. Planing across the grain is more difficult and can easily split the wood. Acrylic can be planed but the process blunts the blade quickly and separate planes should be kept for this work.

Shearing

This is the one cutting process that does not involve a toothed cutting action. Scissors, tin snips and a guillotine all rely on two cutting edges passing by each other, removing the waste without swarf (metal shavings and chippings), dust or filings.

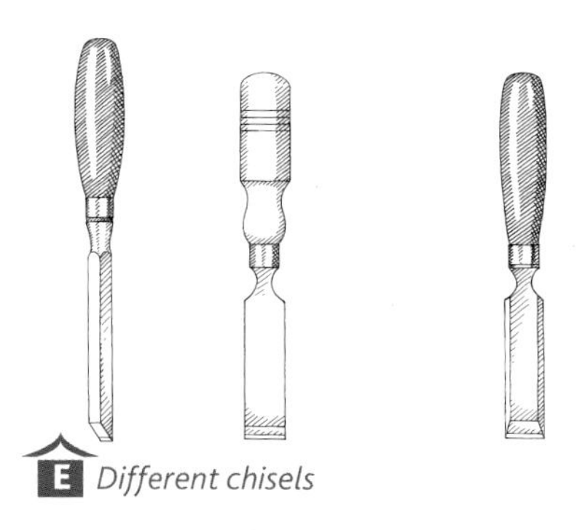

E *Different chisels*

Drilling

Apart from simply making holes in material, drilling can be a quick method of removing quantities of waste material. By choosing the relevant drill bit and holding device all resistant material can be drilled. Although drill bits vary enormously in shape, size and material they all work by rotating a cutting edge clockwise against a material. Unwanted materials are removed by certain features on the drill bit.

Holding

There are many tools designed to hold material securely and safely (Picture E). They allow you to concentrate on working with tools with greater control and to avoid coming into contact with sharp cutting edges. It is very important to use holding tools to ensure your safety.

Activity

1. Sketch a variety of hand tools you have used in your school workshop. Clearly label each part of the tool and explain how the tool should be used correctly.
2. Research how a number of hand tools are maintained and then produce a table to show your results.

Summary

Hand tools should be well maintained to ensure effectiveness when being used for cutting and shaping.

Hand tools can be difficult to use and take patience and practice to be used accurately.

Hand tools present a number of safety issues to the user and therefore materials should always be held securely using an appropriate holding device.

10.2 Power tools

Using power tools can often make working materials easier and more accurate. Hand-held power tools are often battery operated and therefore are increasingly available in school workshops because the risk of electrocution to the user is minimised.

Objectives

Have a working knowledge of a variety of power tools and the particular processes that they relate to.

Be able to select hand or machine power tools and relate them to a particular process.

Machine tools

There are a variety of **machine tools** in the school workshop.

Key terms

Machine tools: larger machines in a school workshop that are normally fixed to either a bench or the floor are more powerful and should have a range of safety devices to protect the user.

Pillar drill

The pillar, or pedestal, drill (Photo **A**) can be used on most materials and the handle offers a mechanical advantage so less effort is required to drill materials. Materials can be clamped to the table, which can be positioned and locked in place, increasing the accuracy of any holes being drilled. The use of the depth-stop and a chuck holding the drill bit in a fixed position also improve precision.

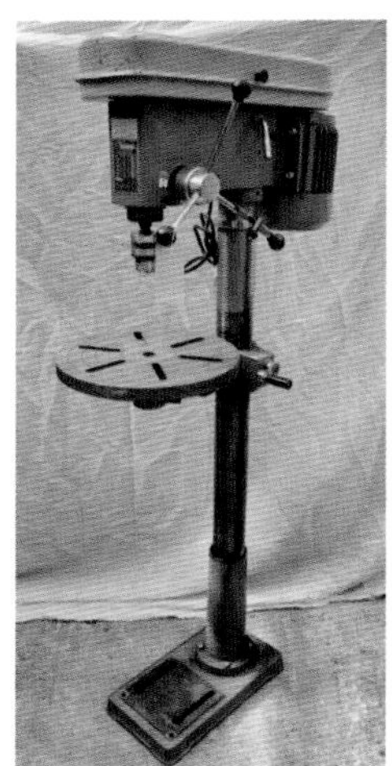

A *A pillar drill*

Belt sander and disc sander

A belt sander (sometimes called a linisher) can be used for trimming accurately to a line used for marking out or for making material flat. A disc sander is more commonly available and can be used for accurately removing unwanted material (especially end grain). Both machines should have guards and dust extraction to protect the users, and be securely mounted. Many belt sanders can be either used in a vertical or horizontal position.

B *Disc sander*

Wood lathe/metal cutting lathe

Both types of machine use a turning action to shape material that is held securely and rotated, as a turning or cutting tool is brought into contact with the material. The wood lathe has traditionally been used for producing round and tapered shapes used for chair and table legs. The metal cutting lathes can be used for more precise work, such as creating conical shapes as well as cutting screw threads.

Band saw

Specialist blades are available for cutting most materials. Blades are one continuous band of flexible steel and are available in different sizes. An adjustable fence allows materials to be cut repeatedly to a chosen length. Some smaller versions of the band saw are used by students under close supervision in some workshops.

Scroll saw

A scroll saw (Photo **C**) is used to cut intricate curves and is sometimes referred to by its trade name, Hegner saw. Scroll saws use blades similar to coping saws and operate through a quick reciprocal (up and down) motion. Different blades are available for a variety of materials and tasks. Some types of this saw are portable but it is important that they are always held down securely.

Milling machine

Milling machines use a rotating multi-toothed cutter to shape materials using a high level of precision. The important feature is that the machine bed is capable of being moved in three separate directions. The x-axis is when the machine bed is moved backwards and forwards; from left to right is called y-axis; and the z-axis is when it is raised or lowered.

Mortising machine

The mortising machine is a less common machine in schools but is a quick and accurate method of repeatedly cutting mortise joints. The machine uses a drill bit held inside a specialist square and hollow chisel. The drill bit clears out most of the material, and the chisel ensures the edges are straight and clean.

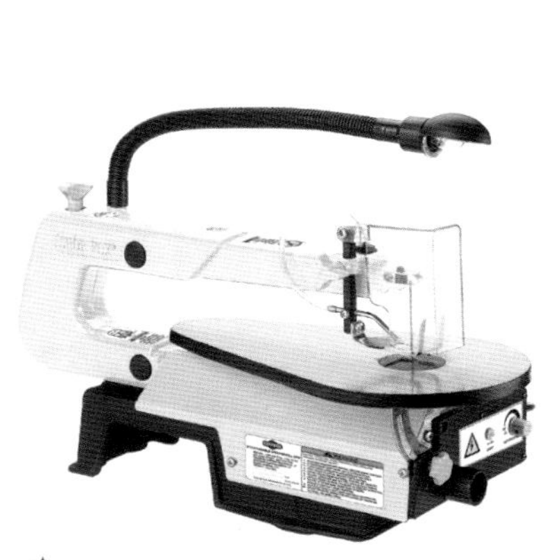

C A scroll saw

D Hollow mortise chisel

Hand-held power tools

Cordless drill

Modern cordless drills (Photo **E**) use low-voltage rechargeable batteries and offer a safe alternative to the traditional hand drill. Most have a clutch setting that also allows them to be used as a screwdriver. They can be easy to use but the weight of the battery can sometimes make it difficult to hold and achieve accurate results. Many drills now available are designed more ergonomically with better balance and soft-grip handles. They can be used away from a power supply without the problems of trailing wires but they have less power than an equivalent mains-operated drill.

E A cordless drill

links

The traditional milling machine has been improved by adding specialist motors to each axis that can be controlled by a computer. See the CAM section on pages 94–95.

Activity

1. Sketch two machine tools that are in your school workshop. Clearly label any parts of the machine you think require regular maintenance. Label any special safety features.

Remember

- Many powered, portable hand tools can present safety issues and great care and thought should be given to their use. Cordless low-voltage tools offer the safest option.
- As with all drilling and cutting work, the material should always be clamped safely and securely.

links

See page 82, The designer, for further details about wood turning.

Biscuit jointer

A biscuit jointer provides a quick and easy way of joining pieces of wood together. They work by cutting crescent shaped slots into both items to be joined and then bringing the two pieces together around a tight fitting, flat and elliptical piece of wood creating a joint. The slots are cut using a blade resembling a thick circular saw blade that is plunged into the wood to a depth set by the size of the biscuit to be used. Some cutters, for safety, feature a slow start motor that gradually builds up the speed of the motor in order to prevent any kick upon switch on.

F *A biscuit jointer*

Jigsaw

Jigsaws can be used for making straight cuts or for curved shapes. There are various types of blade available, which are suitable for most materials. The work must be held securely because the cutting action causes vibrations. Cutting can often be slow and inaccurate on some materials.

Palm sander

There is a wide variety of palm sanders available. Many of them have a quick-change facility for the abrasive pad that comes in various sizes and grades enabling access to even the smallest shapes. They are ergonomically designed to fit into the user's hand and often come with soft-grip handles to increase both comfort and grip. The material being sanded should always be clamped firmly to a bench.

G *A palm sander*

Router

Routers are increasingly popular and are used mainly with wood. With the help of a guide they can be hand-held to produce slots, to cut out shapes following a template or to produce an edge decoration on wood, called a moulding.

The router can be mounted in a specialist table that allows the machine to act more like a spindle moulder. Instead of having to clamp down the piece of wood and work along it with a hand router, you feed the wood past the cutter.

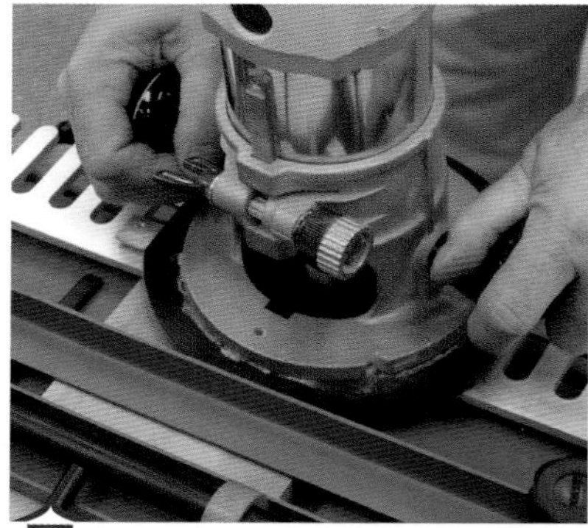

H *A router*

Activities

2 Research some modern hand-held power tools.

3 Use your research to show how improvements in design have made these tools more versatile and more comfortable to use.

Summary

Machine tools are powerful and need special safety arrangements to protect the user.

Both machine tools and hand-held power tools can offer greater speed and accuracy when processing material.

Hand-held power tools can be faster and take less effort but they may be less accurate if used without the aid of guides etc.

AQA *Examiner's tip*

- You will need to know about a number of machine and hand-held power tools, and understand why they are selected for specific manufacturing processes.
- You should be able to fully explain any manufacturing advantages or disadvantages that machine tools have in a school workshop.

11 Techniques and processes

11.1 Marking out

When you begin your making you must consider marking out carefully as it is very important if you are to work accurately and avoid wasting material. The majority of marking out is carried out by placing lines directly onto the surface of the material using traditional tools or CAM equipment (laser cutters/engravers). It is important with either method that you work from a **datum edge** or **line** to ensure components are accurate and within expected **tolerance**.

Objectives

- Have a working knowledge of marking out tools.
- Be able to describe a variety of marking out procedures.

Marking tools

Choice of marking tools is usually determined by the material. Certain marking tools leave marks on the surface and can normally be removed, while some tools actually cut or score the surface and these marks are more difficult to remove.

A *Marking tools*

Name of tool	Notes	Material
Pencil	This is the best tool to be used on a variety of materials, even acrylic if the protective paper is left on. A softer pencil (2B) is easily rubbed out, but for thin, crisp lines use a harder pencil (H).	Wood, acrylic, paper, card
Chinograph pencil	Lines can be easily removed but lack crispness.	Plastics
Marking knife	These make a small cut into the material and can give a clean edge when sawing or chiselling across the grain. Mistakes can be difficult to remove.	Wood, card
Scriber	Used when a very thin, accurate line is required. To help see the line the material is normally coated with engineer's blue.	Metals
Spirit pen	Markers or spirit pens (permanent) come in different thicknesses. Thin pens should be used where accuracy is required. Aluminium sheet is often marked out using a marker as it can be removed by solvents.	Plastics, aluminium

Straight lines

Straight lines that are placed onto materials will be parallel to, at an angel to, or square (90°) to, an edge.

Curves and circles

Marking circles and curves accurately is dependent on the tool(s) you are using not moving or slipping on the surface. When carrying out your own making, check the advice below.

Templates

Templates and stencils are normally used to mark out odd or complex shapes, particularly if the process is to be repeated several times.

Key terms

Datum edge/line: this is used to make all measurements from. It ensures accuracy and prevents errors accumulating.

Tolerance: the amount of error that can be allowed and is often given as + mm. It is more important in some products than others.

The material that is used to make a template will depend on the number of times it will be used. Templates are very popular in manufacturing industries involved in volume production. They are also very useful in school for improving accuracy of low-level manufacture. Templates can successfully be produced from using print-outs (actual size) of CAD drawings.

B *Tools for making straight lines*

Name of tool	Notes	Material
Ruler	These are available in different lengths but are normally 300 mm long in workshops and have a zero end, which makes it easier to measure accurately and 'set' a marking gauge.	Wood, metal, plastic, card, paper
Try square	Used to mark right angles to an edge. It can also be used to test if surfaces are flat and true. Engineer's square has the same function but is smaller and is all made from steel.	Wood, plastic, metal
Adjustable bevel	Can be adjusted to mark lines at different angles. Can sometimes be part of a more complex combination square that allows a range of marking out procedures in a single tool.	Wood, plastic, metal
Mitre square	Used to measure 45° angles.	Wood, plastic, metal
Marking gauge	Used to mark parallel lines on wood. It is adjustable and is used along the grain. Mortise or cutting versions are available.	Wood
Odd-leg callipers	These have a similar function to a marking gauge. They are used to mark parallel lines with the stepped foot held against the edge of the material.	Metal, plastic

C *Tools for making curves and circles*

Name of tool	Notes	Material
Spring dividers	Used to produce very accurate circles and arcs as they have a screw for adjustment. A small indent in the material prevents the point from sliding.	Metals, plastics
Compass	Different versions allow pencils, pens or spirit markers to be held.	Wood, paper, card, plastics
Centre punch	Ideal for making a 'dot' for locating a drill or compass/dividers point on metals. For plastic, a piece of masking tape placed on the surface provides better grip for the compass point.	Metals, some plastics

D *Tools for making straight lines*

Precision marking out

When precision is required and tolerances are small, it will be necessary to mark out using a surface plate, surface gauges, vee blocks and angle brackets. The surface plate is made from cast iron and very accurately finished. The surface should be protected with a cover and lightly oiled to prevent damage. A ruler is not always suitable for precise measurement so more accurate measuring devices are the vernier calliper (Photo **E**) and micrometer (Photo **F**). They are designed to provide a high degree of accuracy (0.02 mm and 0.01 mm respectively). Increasingly, electronic versions are available.

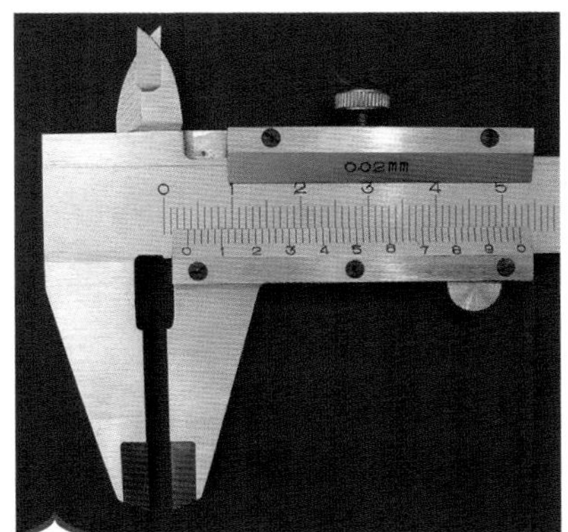

E *Vernier calliper*

F *Micrometer*

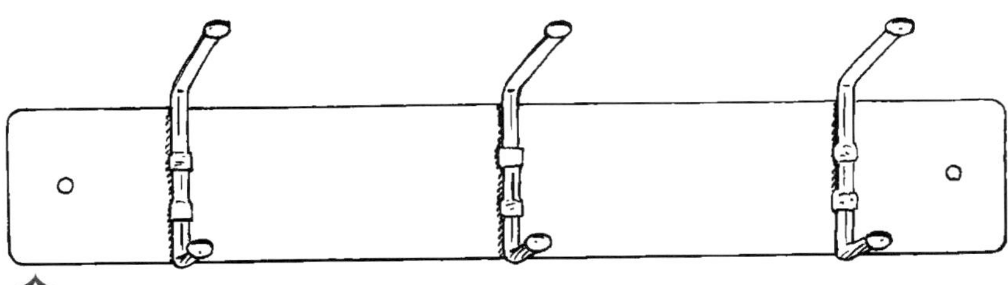

G *Metal coat hook*

Activities

1. Using notes and sketches, show clearly the four stages of marking and squaring a line around a piece of wood. Name the tools used in the process.
2. Study the picture of a metal coat hook in Diagram **G**. Produce a series of sketches to show how the product would be marked out on a sheet of aluminium.

Summary

Careful marking out with the correct tool, machine or template is important in achieving a high degree of accuracy in your making.

Planning your marking out can save wasting both material and your time. *Measure twice; cut once* is a golden rule.

Precision and a high degree of tolerance is achieved by the use of specialist marking and measuring tools as well as CAM equipment.

AQA Examiner's tip

- You will need to learn the importance of marking work out accurately, selecting the correct tools and using them effectively.
- You should also be able to correctly name each marking out tool and, using notes and sketches, fully describe its operation.

11.2 Joining wood

Box joints

Many wooden products of all shapes and sizes are made in some form of a box or carcase that may be made using a simple mitre joint or a more complex **wood joint**. Smaller items such as pencil or jewellery boxes are often made from solid wood. Larger items such as bookcases and cupboards are made in manufactured boards. Industrial manufacturing techniques are increasingly using KD fittings rather than the more traditional jointing techniques.

Butt joint and lap joint

These forms of joint are relatively quick and easy to make but lack strength. Butt and lap joints can have their strength improved by the addition of dowels or biscuits (an oval-shaped, highly dried and compressed piece of wood, usually beech). The biscuit is placed in the slots made by the cutter and the wood is clamped together. The wet glue expands the biscuit, further improving a bond that is often stronger than the wood itself. Without dowels or biscuits a butt-jointed box will often rely on some form of reinforcing in the box, such as a manufactured board held in a rebate.

Comb joint

Comb joints, sometimes called finger joints, are strong because they interlock and there is a large gluing area. They can be used for both natural and manufactured boards. They are widely used in the industrial production of furniture as they can be quickly and accurately cut using machines. They have almost replaced the dovetail joint, which is traditionally a hand-cut decorative joint, and is far more difficult to reproduce using machinery.

Partition joints

A box construction often needs to be a split into a number of useful sections, or partitions. Boxes are normally divided up using one of three methods. A butt joint is the simplest and requires the use of either nails, or dowels and glue. A housing joint is stronger but can be difficult to create accurately and is also glued. The most difficult form of this type of joint is the stopped housing, which gives the neatest appearance.

Objectives

To have a working knowledge of joining wood using several suitable jointing methods.

Key terms

Wood joint: a system used to join two or more pieces of wood together that might have some mechanical strength and is likely to be permanently joined using an adhesive.

links

See KD fittings and fixings on page 28 for more information.

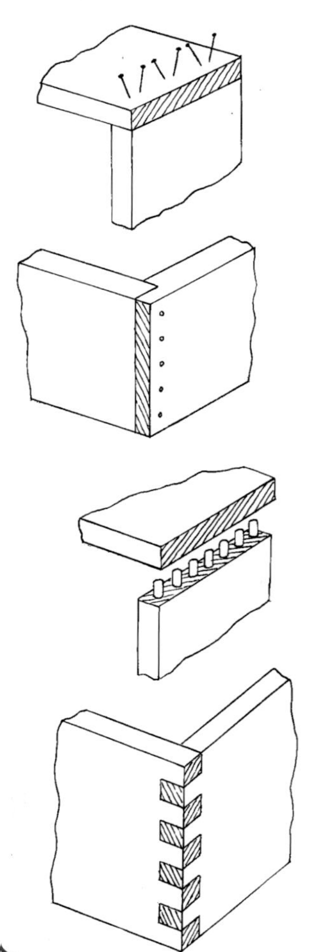

A *Butt, lap, dowel and comb joints*

Frame joints

Frames are normally made using natural timbers and are commonly used in manufacturing stools, chairs or doors. The most traditional joint is the mortice and tenon, which has a mechanical strength even before it is glued. However, this is increasingly being replaced by the dowel joint due to the availability of drilling jigs and its wide use by industrial furniture manufacturers.

Halving joint

The halving joint is the simplest form of frame joint. It has no real mechanical strength and relies on glues, screws or dowels for strength but is helpful however for locating the materials for assembly.

Mortise and tenon joint

The mortise and tenon joint is usually used to form a right angle joint that is a very strong and neat joint when glued. There are other varieties of mortice and tenon available that may not form a right angle or have been selected for their strength, appearance and the location within the frame.

Dowel joint

The dowel joint is also very strong and allows wood to be butt-jointed. If you use a dowel joint in the school workshop, drilling the holes accurately is the most difficult part and the use of some form of drilling jig is recommended.

Activity

Look carefully at the products in the photographs. Describe fully the materials and tools used to cut two different types of wood joint that the manufacturers have used in the production of these products (Photo **C**).

C *Examples of wood products*

Summary

Natural wood and manufactured boards are often joined using different techniques.

Wood joints can be used to create a feature that enhances the appearance of a product.

links

See Quantity production on page 130 for more information.

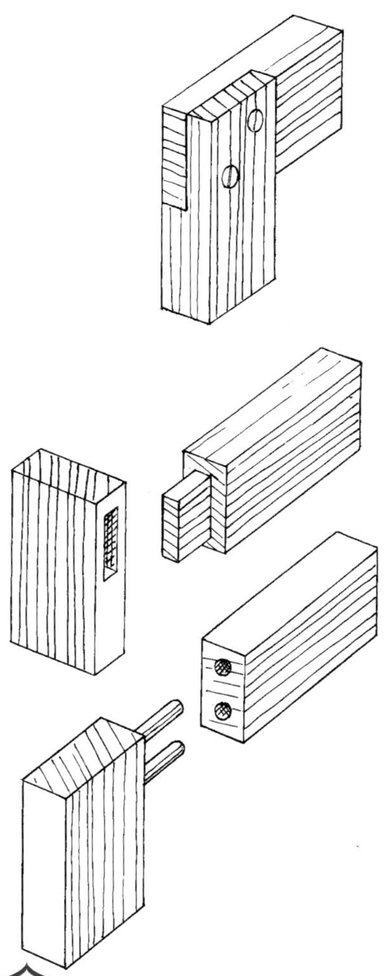

B *Halving, mortise and tenon, and dowel joints*

AQA Examiner's tip

- You should be able to name a number of wood joints and give reasons for their selection.
- You should also be able to explain fully, using a series of notes and sketches, the procedure for cutting selected wood joints.

11.3 Joining metals

The most common way of joining metal to itself or another metal permanently is to use a process that involves heat. The heat processes that are commonly available in schools are soft and hard soldering. Another process is welding and some school workshops may have oxy-acetylene or electric arc-welding equipment. There are, in fact, many different industrial ways of welding metals.

Objectives

Have a working knowledge of the processes that can be used to join metals permanently.

Remember

Due to the temperatures required for joining metals, great care should be taken when handling all metals. The specialist tools available such as heat resistant safety gloves, long handled tongs or pliers should be used to handle hot metals.

Soft soldering

This is used as a quick method of joining copper, brass and tinplate when little strength is required in the joint. It is also used for fixing electronic components into a circuit. The **filler rod** or solder melts at a relatively low temperature and traditionally was a mixture of tin and lead but due to the health risks of using lead has now been replaced by an alloy of tin, silver and copper. To make the joint as strong as possible it must be thoroughly cleaned using wire wool or emery cloth. The **flux** that is applied prevents oxidisation but is corrosive so should be cleaned off after the joint has been made. 'Multicore' solder used for electronic components contains fine lines of flux running along its length.

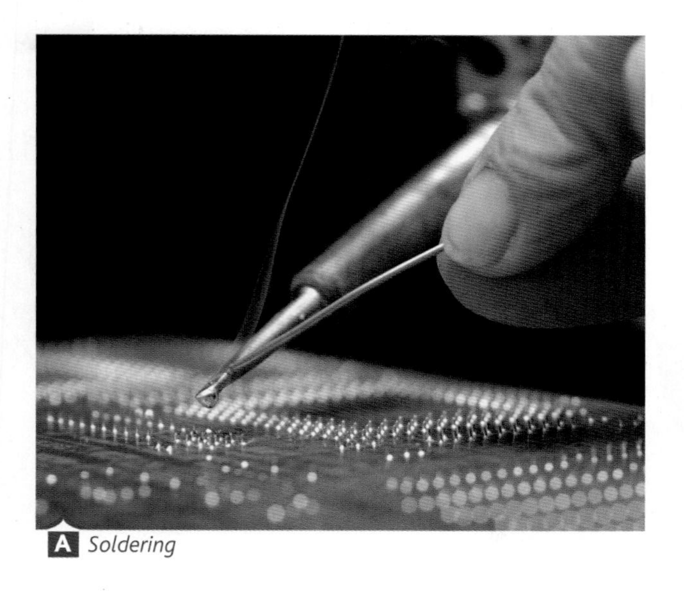

A *Soldering*

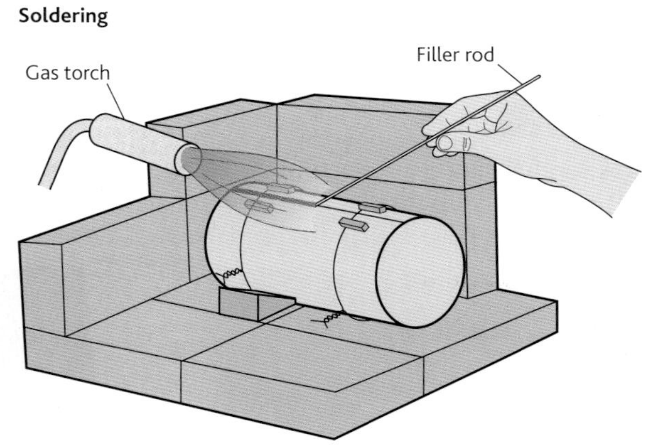

Hard soldering

Hard soldering, also known as silver soldering, uses a filler rod that is an alloy of silver mixed with copper and zinc and melts at temperatures between 600°C to 800°C. This range of temperatures enables work to be joined in several stages with the highest melting point being used first through to the lowest melting point solder called 'easy-flo'. This prevents earlier joints coming apart when applying heat for later joints.

Brazing is the same process as silver soldering but is used for joining mild steels and uses filler rods with melting points in the range 870°C to 880°C. A brazing torch using gas and air or an oxy-acetylene welding torch needs to be used to fabricate these joints. To successfully hard

Key terms

Filler rod: an alloy that is used to bond two metals together as they are heated.

Flux: a chemical cleaner used to clean metal as heat is applied.

solder, the joint must be prepared by thoroughly cleaning with emery cloth and then applying flux to the joint surfaces.

Welding

Many schools have access to the basic methods of welding: oxy-acetylene and electric arc. Both of these fuse steels together to produce a very strong joint. Oxy-acetylene welding uses acetylene burnt in oxygen to produce a flame at approximately 2,500°C. Using the heat of the flame on the joining edges melts the metals and a filler rod is introduced to help fuse the materials together by melting into a pool that sets on cooling. In electric arc welding the heat required to melt the metals is provided by a current passing through a gap (arc) between the filler rod (electrode) and the metal. The electrode is coated in a flux to prevent the joint becoming oxidised.

Industrially there are a much wider range of methods of welding such as MIG (metal inert gas) and TIG (tungsten inert gas). These forms of welding have been refined to produce quicker and more reliable production methods. Spot and seam welding use the electric arc process and are popular methods of joining metals industrially as well as for the robotic assembly of car panels.

B *Joining metal using oxy-acetylene*

C *Guitar stand*

AQA Examiner's tip

- You should be able to sketch and describe a variety of methods of joining metals.
- You must learn about the different jointing methods for metals and be able to explain why particular methods are selected and used to manufacture metal products.

links

See Mechanical methods of joining materials on page 34–35, to find out about joining metals with bolts, rivets and pop rivets.

Activity

Look at the photo of a guitar stand (Picture C). It has been made with a number of bent mild steel tubes that have been joined by brazing.

Produce a list of helpful hints that would allow somebody to carry out the brazing procedure successfully in a school workshop.

Summary

A variety of metals can be joined using different processes, all of which involve the application of varying degrees of heat.

Preparation is required to successfully join metals using the soldering process.

Filler rods are available that melt at different temperatures and can be used to either fuse or 'glue' the metals together.

Industrially, a number of welding processes have been developed to increase speed and accuracy in production.

11.4 Casting

Casting is a process that involves pouring molten material into a shaped **mould** and is used to produce a range of shapes that would normally be difficult to fabricate or machine from a single piece of material without the process wasting both materials and time. Materials such as plastics and concrete can be cast but the process is particularly suited to metals. Traditionally, casting in schools has used aluminium alloy (for example, LM4) as it is readily available and melts at temperatures around 750°C. Pewter casting is increasingly popular in schools due to its lower melting point and the need for less sophisticated equipment. The casting process ensures reduced waste (offcuts can be re-melted and re-used), as well as enhanced product strength – a result of the internal structure changing as cooling/solidification takes place.

Casting using aluminium

Sand moulds are commonly used both in school and in industry to produce complex metal shaped castings. The process of casting involves a number of stages – **pattern** making, mould making, melting and pouring and finishing (fettling).

The simplest form of pattern is a flat backed pattern, which is detailed on one side only and used for such things as decorative plaques. More complex shapes are made using a split pattern, which has two halves that locate together with pegs. It is important when making your own pattern to ensure that sharp corners are rounded, internal corners radiused and vertical surfaces tapered to create 'casting draft', all of which help the withdrawal of the pattern from the sand (Photo **A**).

To produce the mould ready for pouring a standard two-part moulding box with locating pegs is filled with an oil based moulding sand (for example, petrobond/petrabond). The stages of making a moulding box can be seen in Table **B**.

Once the metal has solidified and cooled the sand is broken away to be reused and the metal shape is fettled by removing the runner and riser, and where necessary machined to size or to remove surface imperfections. The metal is then ready for an appropriate finishing process.

Pewter casting

Pewter is an alloy with a melting point (approximately 230°C) that is low enough to allow a small gas/air torch or heat gun such as a paint stripper to provide more than enough heat to melt and maintain a pool of molten pewter ready for casting. As with aluminium casting there is no wastage as all unused pewter can be re-used.

In schools moulds can be made from a range of materials including medium-density fibreboard (MDF), necuron foam (grade M-651-2) or acrylic. All these materials are ideal for making moulds but they do degrade over time, although many have been re-used up to 50 times. The pattern/cavity can be cut in the material either using hand tools or, for more accurate results, or CAM machine. The pattern/cavity is then

Objectives

Have a working knowledge of at least one casting process.

Key terms

Mould: a container and packing material that is used to create the space/shape required for pouring in the liquid material.

Pattern: an exact copy of the shape required. Certain shapes are made slightly larger than the required finished size to allow for shrinkage as the material solidifies.

A *Flat back pattern showing the required 'draft' to help removal from the mould*

Remember

Great care is required when pouring molten aluminium and the correct protective clothing must be used. Aluminium should be heated to about 750°C. Pouring into the runner should be slow but continuous until metal appears in the riser, indicating that the mould is full.

sandwiched and clamped together with two other sheets of material. Cuttlefish bone, which is a natural, soft material, has traditionally been used in schools for this casting process.

B *Stages of making a mould*

Stage 1

The pattern is placed on a baseboard and the 'drag' (the bottom half of the moulding box) is placed over it.

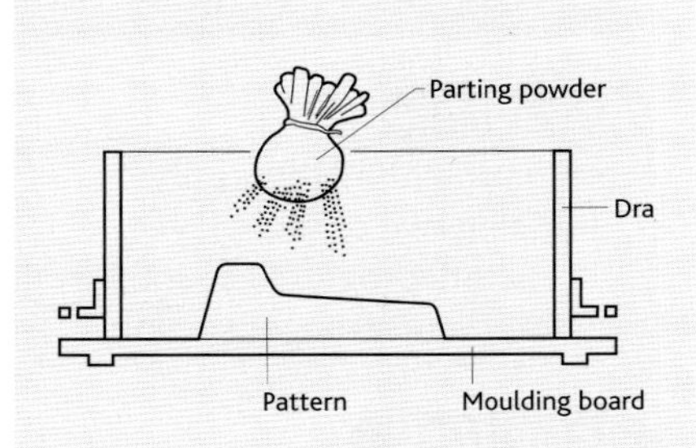

Stage 2

The sand is sieved to remove lumps and then is very tightly packed into the drag.

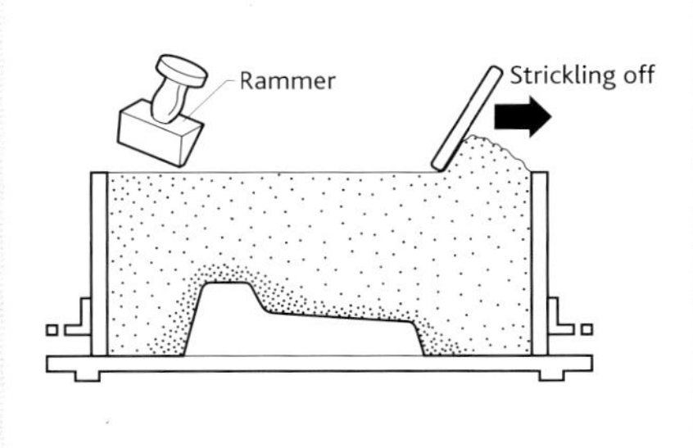

Stage 3

The cope (the top half of the moulding box) is then fitted together with the drag using locating pins for correct alignment.

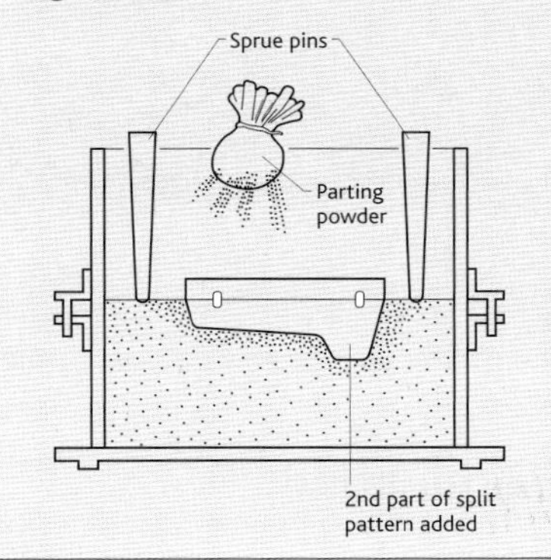

Stage 4

Tapered pins are introduced to create runner/ riser. Sieved sand is tightly packed into the cope.

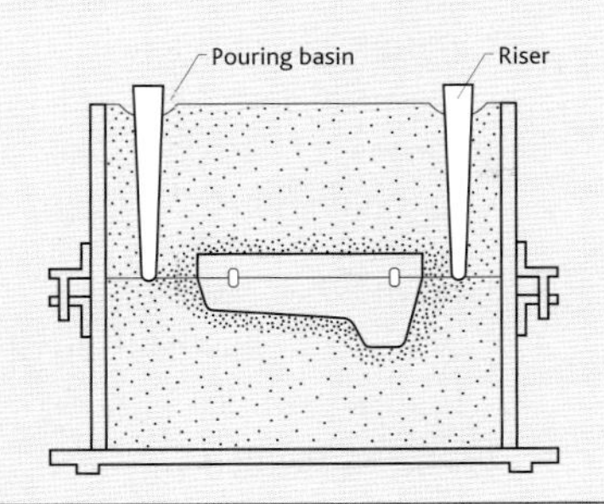

Stage 5

The cope is carefully removed and gates or channels are then cut from the patter to the runner/riser. The pattern is removed and loose sand is gently blown away. The cope is refitted. The mould is now ready for pouring.

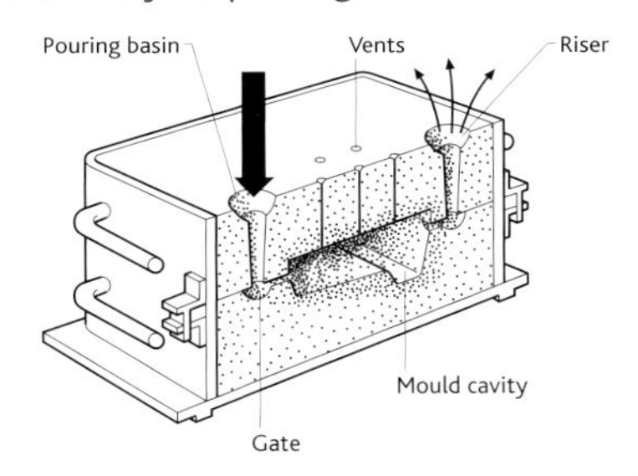

C *An MDF mould for pewter casting*

Activities

1. Using notes and sketches fully explain how pewter jewellery might be manufactured in a school workshop.
2. Research one other casting process (lost wax, resin, die or investment) and produce an information sheet (A3) that could be used to easily and clearly explain the process to other pupils in your group.

Summary

The casting process can be used to manufacture complex shapes in a number of materials.

Metal casting can produce very strong and complex shapes that would be difficult to make using other manufacturing techniques.

Waste offcuts are easily recycled.

AQA Examiner's tip

- You should be able to select and sketch the appropriate casting method that is used to produce different metal shapes. It is important that you can show the important parts of the process and in the correct order.
- You should be able to describe in detail why casting is important in the production of complex shapes.

11.5 Forming woods

A *Laminated roof beam structure ice rink*

Objectives

Have a working knowledge of the wood laminating process.

Be aware that wood can be formed into a variety of shapes using a turning process.

Key terms

Laminating: the process of bonding two or more layers of material together to form a thicker and stronger section.

Laminating

Apart from creating material that is often longer or thicker than what is available in one piece, for example, plywood, **laminating** is also used to produce shaped materials with improved properties or to obtain shapes that cannot be cut from one piece of material (Photo **A**). Thin strips of wood are deformed or bent into curved shapes (clamped against a former) with each layer of wood having their adjoining surfaces glued together. Once the glue has dried the bent shape is maintained as each laminate is no longer free to move. The formed piece of wood due to the grain pattern created has increased mechanical strength and offers some interesting aesthetic opportunities in both edge treatment (different coloured layers) and the possibility of using more expensive decorative veneers on the outside surfaces of the laminated shape. Different glues can be applied dependent on where the laminated structure is to be used.

Formers

To produce laminated shapes in a school workshop there are a number of ways of using formers to create the desired bent shapes. The simplest method is to use a solid wood former, which is suitable for small or simple shapes (Diagram **B**). The surfaces of the former must be smooth and there must be an allowance for the thickness of the laminates. The former can be lined with rubber or cork to help with slight irregularities in both the former surface and laminate thicknesses. For larger and more complex laminated structures a built-up wooden former can be used. These are built up from different sections of wood and for many bent shapes there is no need to make accurately matched male and female formers. The glued laminated shape is clamped to the former using thick strips of wood sash clamps, and G clamps. With flexible formers there is a solid male part to the formers and the female part of the former is usually a spring steel band that is used to clamp the work against the male section. This method can save preparation time but is limited to curves being formed in one direction only.

Activities

1. Using notes and sketches fully explain the process of manufacturing the chair frame (Photo **D**) in a school workshop.
2. Identify a number of items that are made by laminating. Carefully count the number of layers in each item and the thickness of each layer. Record your findings in table form.
3. Describe how you would prepare a square section of wood to produce a round table leg using a wood turning process. Use notes and sketches.

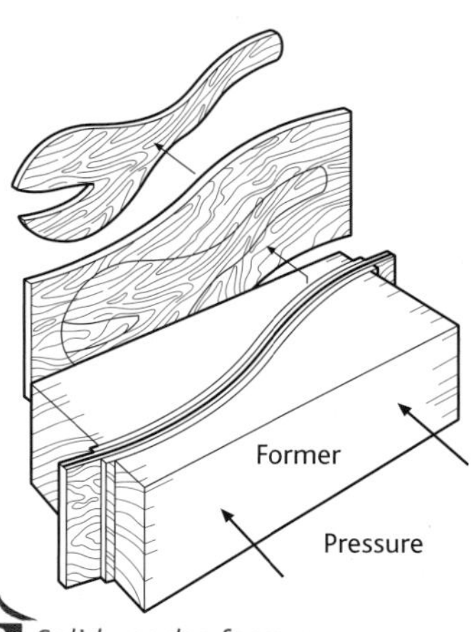

B *Solid wooden former*

Wood turning

The wood lathe traditionally has been used for **wood turning** to produce wooden objects such as chair legs, bowls, lamps and patterns for either casting or vacuum forming. Turning between centres has the work supported at both ends with the driving forked centre located in the headstock and a dead centre (possibly revolving) located in the tailstock. This method is used to produce long pieces of work. Face plate turning is used for such things as bowls, dishes, formers and bases and involves a face plate being used on either the left- or right-hand side of the headstock, dependent on size. The lathe provides the rotation of the material and the cutting is carried out by a range of specialist tools. They all have long handles to give additional leverage so as to withstand the turning forces. There are broadly three types: gouges, skew chisels and scrapers. Gouges and skew chisels use a cutting action and need to be kept sharp. The gouge is used for rough shaping or truing-up and the skew chisel is more suited to final shaping and finishing. The skew chisel is the most difficult to use but in experienced hands can produce an excellent finish. Scrapers also need to be kept sharp and are the easiest to use but the work they produce will require sanding. Tools can vary in shape and size and their selection is dependent on each stage of turning and the size of the wood being turned.

D *Laminated chair*

Summary

Laminating is used to produce mechanically strong wood products that are not normally available from a single piece of wood.

Laminating wood uses formers, clamping and adhesives.

Wood turning can produce a wide variety of shapes in wood.

Specialist tools are used for turning wooden items on a lathe.

Key terms

Wood turning: a manufacturing method that uses specialist tools to shape wood that is being spun or rotated.

C *Wood turning lathe*

Remember

- Wood used for wood turning should be free from splits.
- Wood should be prepared by cutting off waste/planing edges.
- Work should be secured to face plate with screws long enough to hold the work securely but not so long as to break through the work.
- For turning between centres the work should have a saw cut to locate the forked centres.
- The tool rest should be as close to the work as possible. Check by turning the work by hand.
- Keep the tools sharp.
- Hold the tool firmly, using the long handle for leverage to prevent accidental 'snatching'.

AQA Examiner's tip

- You should be able to select and sketch an appropriate laminating method.
- You should be able to describe in detail why laminating can produce wooden articles that are mechanically strong.

11.6 Deforming metals

Forming or **forging** shapes in metal is often considered to be both quicker than machining shapes from solid metal as well as good economical use of materials. Most forging processes are carried out hot to avoid the risk of work hardening and to ensure they are easier to work into the required shape. Forging is a highly skilled process that has been used for centuries to create decorative products but in industry it is now a sophisticated computer controlled process.

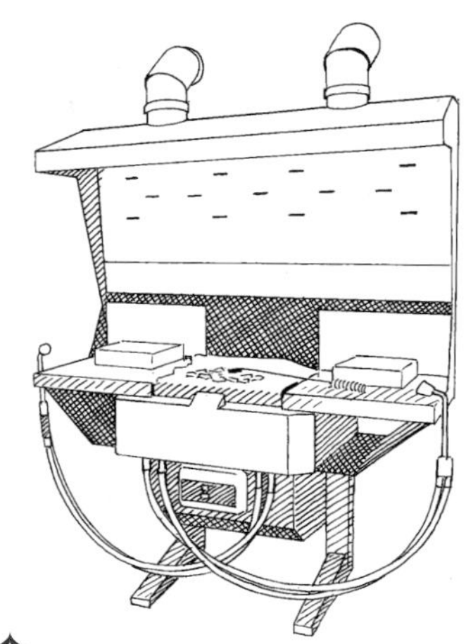

A *Double brazing hearth and chip forge*

Objectives

Have a working knowledge of how to deform metals.

Hot forming metals

Hot forging is often considered to produce very strong components due to the structure of the material being refined that can increase toughness. The most common metals hot formed are mild steel and tool steel. It is important that heat is applied in the correct area and to the temperature that allows deformation. Basic processes are carried out by hand using a hammer, anvil and different stages (Photo **B**). Larger forces can be applied by either a falling hammer or a powered hammer to do the hammering (often known as drop forging). Processes include bending, drawing down, punching and drifting and twisting and scrolling.

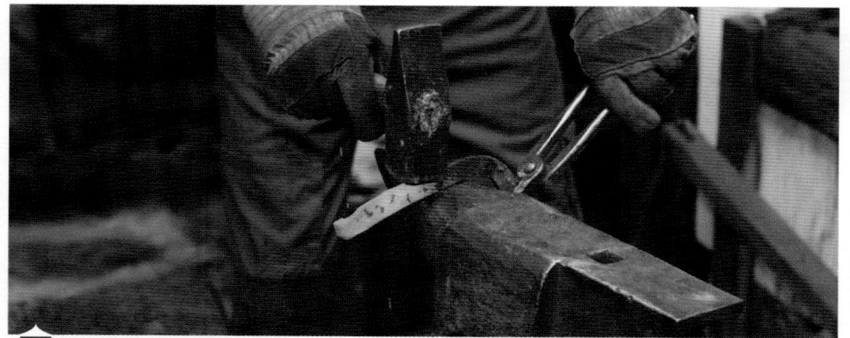

B *Hot forging on an anvil*

Key terms

Forging: a traditional process that uses a hammering action provided either by hand or from a machine to create a variety of shapes.

Work hardening: as a result of deformation of a metal there is an increase in hardness that may eventually cause fractures.

Annealing: the softening of metal by heating to a specific temperature and then allowing to cool.

Cold forming metals

Forming metals cold requires a material that has a high degree of malleability. Copper is used in schools because it is relatively cheap but other suitable metals are silver, brass and aluminium. The constant hammering to the surface causes **work hardening** and therefore needs regular **annealing**. Annealing creates a surface oxide that requires cleaning before further beating. To clean away the oxide you can use either dilute sulphuric acid (potentially dangerous) or a mechanical abrasive (safer) like pumice powder and steel wool. Processes include hollowing, sinking, raising and planishing.

Cold working sheet metals

Folding and rolling sheet materials like mild steel, aluminium, copper and tin plate allow you to produce complex shapes such as cones and cylinders as well as boxes and trays. Before beginning to work with your chosen metal you should practise with card to establish sizes, position of joining flaps and the sequence of bends. Using sheet materials can leave sharp and flexible edges that will need to be bent over to make safe and strong using folding bars.

C *Rolling bars*

D *Box pan sheet metal folder*

Activities

1. The copper bowl (Photo **E**) has been cold formed. Identify other objects made using this process. Sketch the items and name the material used to manufacture them.
2. Research the different long handle tongs used for forging. In a table sketch each type and explain the use of each type.

E *Beaten copper bowl*

Press forming

Many strong everyday products like the components of an iPod, sinks and car panels are produced by pressing thin sheets of metal into a shell shape. The metal sheet is often annealed and then placed between a male and female half of an expensive precision die and then a huge press puts hundreds of tonnes of pressure on the die to create the shape. Problems can occur from this process (tearing and wrinkling) if the deformation is in 3D so the design must avoid both sharp bends and over-stretching the material. Simpler 2D press tooling can be used for the production of channels or ribs and punching holes. Press forming commonly uses different procedures to produce a single component, that is, a car door panel is pressed to shape and then pressed to punch holes in the panel and finally pressed to create a turned edge for strength.

AQA Examiner's tip

- You should be able to sketch methods of forming (both hot and cold) metals.
- You must be able to describe fully the advantages of hot forging in comparison to cold bending of metals.

Summary

Forming or forging is a quick and highly skilled process for deforming metals.

The procedure of forming metals by hot forging prevents work hardening.

Metals can be shaped hot or cold but require a high degree of malleability.

Bending and press forming sheet metals can produce strong sophisticated shapes and panels.

11.7 Moulding plastic

Vacuum formed products

The majority of thermoplastics soften and become pliable at temperatures around 160°C. This makes them relatively easy to deform into complex shapes that would be difficult or impossible to make using fabrication techniques. Unlike the wasting or fabrication process there is no loss or gain of materials when deforming plastics. A number of moulding processes are available industrially but often need complex, expensive **moulds** as they are used for large scale production.

Objectives

Have a working knowledge of how to mould thermoplastics.

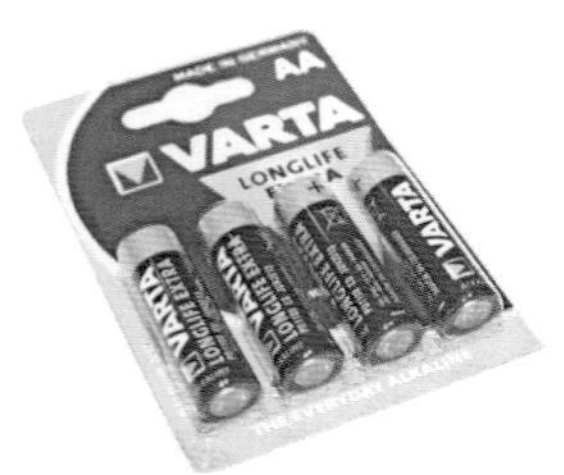

A *Vacuum formed products – a blister pack of batteries and yogurt pots*

Remember

High temperatures are required for moulding plastic and therefore great care should be taken. Heat resistant safety gloves should be used to handle hot plastic in all moulding processes.

Activities

1. Using notes and sketches, fully explain how a plastic food container, such as a yogurt pot, has been vacuum formed.

   Research another moulding technique that could be used to make plastic containers and using your findings produce a flow diagram to explain the manufacturing process.

2. Develop a design that could be used as a simple pen and pencil tray inside a drawer. Using notes and sketches, explain what a cross-section of the mould might look like. Remember to explain key features of your mould.

Key terms

Mould: a 3D template used to create a required shape. It must be made from a material that can withstand both heat and slight pressure.

Vacuum forming

Vacuum forming is one method of forming plastic that is common in schools and is often used to produce items such as trays, cartons, lids, etc. This method of manufacture is used extensively in the packaging industry for batch or mass production of blister packing, food packaging and other small protective containers.

The vacuum forming process works by heating a clamped sheet of thermoplastic until soft. Air is then extracted to create a vacuum, which causes the plastic to be sucked down (helped by atmospheric pressure pushing downwards) onto a mould, sometimes called a 'former'.

The mould that is used must be shaped so that it can be removed easily after the forming has taken place. If you are going to use this process to manufacture a component you should ensure your mould is tapered, has a smooth finish and that any sharp edges have been rounded.

B *Stages of vacuum forming*

Description	Illustration
1 The mould is placed inside the machine and the plastic sheet is clamped to the top of the box using the toggle clamps. The heater is then moved into position to heat the plastic until it softens. The material will begin to sag under its own weight and will appear rubbery.	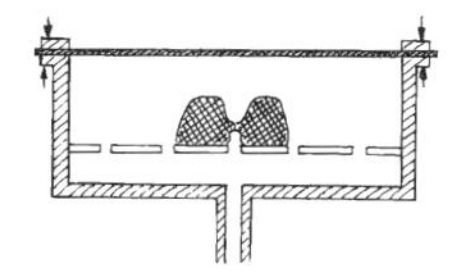
2 The heater is pushed back and the mould, located on the platten, is then raised into the hot plastic before the vacuum pump is turned on.	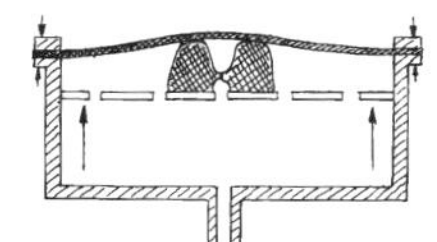
3 The air between the mould and the softened thermoplastic is sucked out by the pump. The plastic (commonly polystyrene) will be forced down over the mould, creating the sharp definition.	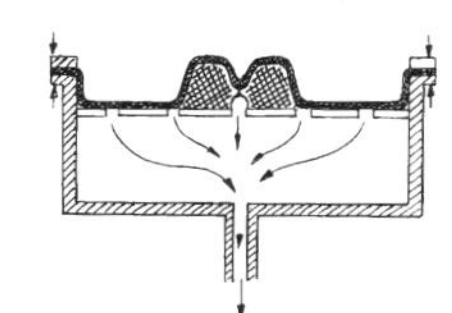
4 The sheet is unclamped from the frame and the mould is removed. Excess material around the moulding is trimmed off.	

Line bending

This is the simplest method of forming thermoplastics. A strip heater that has a narrow opening allows heat to escape in a restricted area, which heats and softens the plastic in a concentrated line. Acrylic is a popular plastic in schools. It can be line bent but care must be taken as it needs to be heated carefully from both sides to prevent blistering from overheating and the possibility of snapping as it is bent.

links

Study Quantity production, pages 130–131, to learn more about jigs.

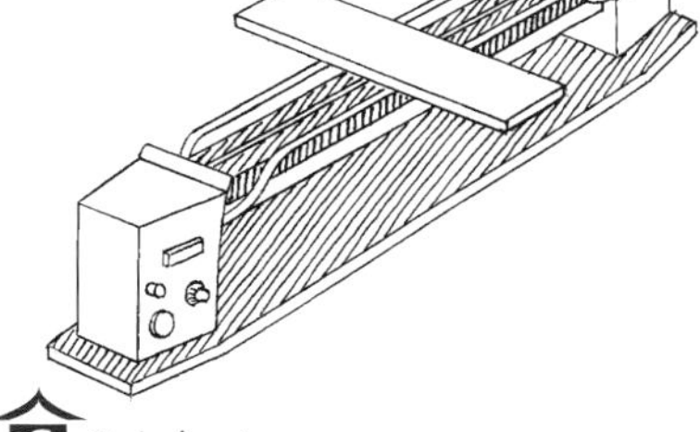

C *Strip heater*

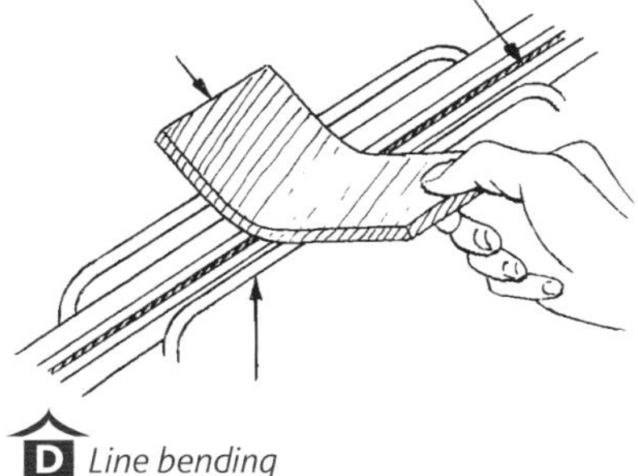

D *Line bending*

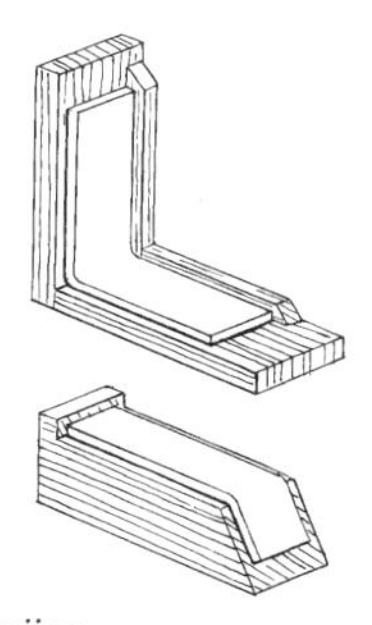

E *Bending jigs*

The use of formers or jigs will help improve accuracy in bending to certain angles and also helps to hold the work while it cools. Jigs are very useful if you have to bend a number of identical components because they help achieve a high degree of accuracy and consistency. Jigs and formers (moulds) should be made from a heat resistant material.

AQA Examiner's tip

- You will need to learn why plastic products are manufactured using certain moulding processes. You should be able to identify a moulding process and give reasons for its selection.
- You should learn and be able to fully describe, stage by stage, at least one moulding process used in a school workshop.

Summary

Moulding thermoplastics is a deforming process that can be used to produce complex, deep-sided products.

Most of the common thermoplastics can be formed in school workshops using the moulding processes described above.

Other, more complex plastic moulding processes exist that are used for large scale manufacture.

11.8 Computer-aided manufacture (CAM)

Computers are now used to control many manufacture machines and have revolutionised industry by allowing the development of products not possible using traditional manufacturing techniques while also having the benefits of cutting time, costs and improving accuracy.

You may have had limited opportunity to use a **CAM** facility but like industry there is an increasing range of CAM machinery that is becoming available in schools. The CAM equipment operates in a way that uses a design programme (for example, Techsoft 2D design) and a computer to produce a set of instructions that control the operation of a manufacturing tool. The instructions are used to control cutting speeds/power and the special electric stepper motors that control the tool path. Some machines work in two dimensions (x and y axis) while some can work in three dimensions (x, y and z axis).

Objectives

Have a working knowledge of several types of computer-aided manufacture (CAM).

Key terms

CAM: computer-aided manufacture-machines that are computer numerically controlled (CNC).

Vinyl cutters

These machines can cut complex shapes from a variety of thin materials and are available in a range of sizes. They are used to carry out work such as card modelling, packaging and sign writing. They only work in 2D but are quick and easy to set up, cutters (expensive) need to be sharp and thicker card requires a number of passes by the cutter. Their relatively low cost means they are often the introductory CAM machine for schools.

A *Vinyl cutter*

3D router

A 3D router is used for accurately manufacturing a wide range of 2D and 3D components in wood, MDF, foam and plastics materials. They are easy to set up and operate and are available with various working envelope sizes (that is, size of material that can be placed in the machine). They require dust extraction to prevent fine dust becoming airborne, clogging up the machinery and causing risk to users. With all of the tooling and equipment required these machines are relatively expensive.

B *3D router*

CNC milling machine

These are similar to 3D routers producing a wide range of 2D and 3D components but often in materials such as steel, brass, alloys and composite materials. Materials are easily held by either a small vice, vacuum or the use of double-sided tape. Different tooling is available for milling, drilling and in some machines scanning. Larger versions with the necessary tooling can be relatively expensive but there are cheaper desktop models available.

C *Modelling/milling machine*

CNC lathe

A CNC lathe accurately manufactures a wide range of complex turned components from steel, brass and alloys, to plastic materials. They are particularly useful for producing quantities of identical pieces. Tooling can be complex and some machines have multi heads that have turning, cutting and parting tools as well as a drilling facility. These machines are relatively expensive.

D CNC lathe with multi head

Laser cutter

Lasers have found a number of innovative applications in industry and smaller systems are now used in schools for cutting and/or engraving the surface of wood and plastic based materials. They are very accurate, easy to operate and are useful for repetitive production. However, their use in the classroom has some health and safety concerns so they should never be left unattended when working (potential fire hazard) and they must have fume extraction. This type of computer controlled machine is relatively expensive.

E Laser cutter

Rapid prototyping systems

Several rapid prototyping systems are available for producing 3D prototype models.

- Stereo-lithography/laser sintering: a laser beam draws each slice of the 2D model on the surface of a bath of liquid resin or fine heat fusible powder. The beam instantaneously cures the resin or melts the powder thus fusing it together. A 3D model is built up from the base. This is a very expensive process.
- Laminated object manufacturing: a model is formed from successive layers of bonded sheet material. Each layer is cut from a sheet of material then bonded to the top of previous layers. This is relatively inexpensive but a slow method.
- Three-dimensional printing: a print/extrusion head similar to a computer ink jet printer builds up layer upon layer of thermo-polymer material, which quickly solidifies to form the physical 3D model. This system has the lowest cost of the rapid prototyping systems.

AQA Examiner's tip

- You should have a working knowledge of a number of methods of CAM.
- You should be aware of the benefits of CAM machinery and the relative cost of the equipment.

Summary

CAM equipment can be expensive but has a number of important advantages when manufacturing.

Special computer software converts designs into instructions that are used to control CAM equipment.

The different types of CAM equipment allow manufacturing in both 2Dand 3D as well as with a variety of materials.

Activity

Create a table that lists the advantages and disadvantages of CAM.

11.9 Quantity production

Dependent on demand, products are made in a range of quantities from large-scale mass production (screws, light bulbs, etc.) to a one-off product (bespoke furniture, your project in the controlled assessment). At each level there are both advantages and disadvantages as seen in Table A. In manufacturing, time is often the most expensive commodity, therefore to increase accuracy, speed of operation and reduce the need for highly skilled workers **jigs** and **templates** are commonly used during production. For jigs to work successfully they must:

- always locate the material against a reference (datum) edge
- avoid waste material getting trapped
- secure the work quickly and easily (hand, toggle clamp or cam)
- be made from suitable materials to avoid wear (likely to lead to inaccuracy).

Objectives

Have a working knowledge of several types of jigs, templates, moulds and formers.

Key terms

Jig: an aid to fast, accurate and repeatable manufacturing operation.

Template: a device to mark out components quickly and accurately.

Scale of production

Your project work may benefit from the use of jigs and templates as their design and manufacture can be a high level task and will add to the complexity and level of accuracy of your work.

Drilling jig

When drilling a number of holes in identical components a drilling jig saves time and increases accuracy. The number of repeated operations

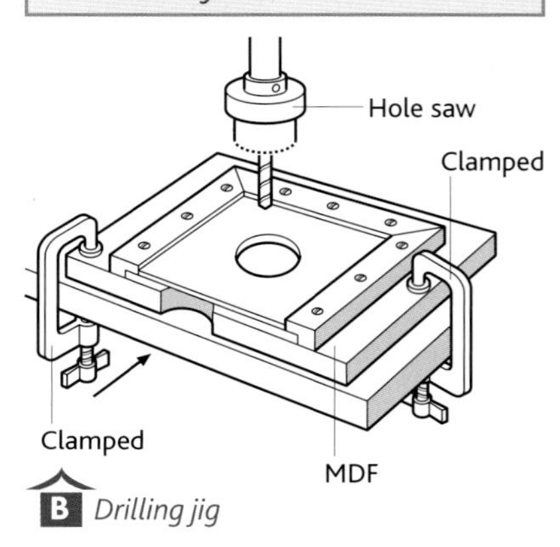

B *Drilling jig*

A *Scale of production*

Level of production	Description	Example	Equipment tooling cost	Labour costs	Skill level	Production costs	Efficiency
One-off	A product that is designed and manufactured for a specific situation. Most products at this level of production are hand made.	A sculpture, your own coursework	↓	↑	↑	↑	↓
Batch	Machinery is set up to manufacture a set number of products. Jigs and tooling are used to manufacture at this level. The same equipment and workforce will be available for producing different 'batches'.	Table, stools					
Mass	A product goes through various stages using dedicated specialist equipment and workers (each with specific responsibilities). Many items mass produced are produced in high volume, for example, plastic milk containers.	Car, light bulbs, nuts, screws, plastic containers					
Continuous	There are very few processes that are in fact continuous but some industries, for example, food, oil and steel have scale of production at this level.	Various food items, steel, petrol, lubricants					

Direction of arrows shows level of increase

often determines what material the jig will be made from. You might make a simple drilling jig from MDF or mild steel to drill a number of identical holes but in industry drilling holes in the hundreds, a jig with hardened steel inserts might be used.

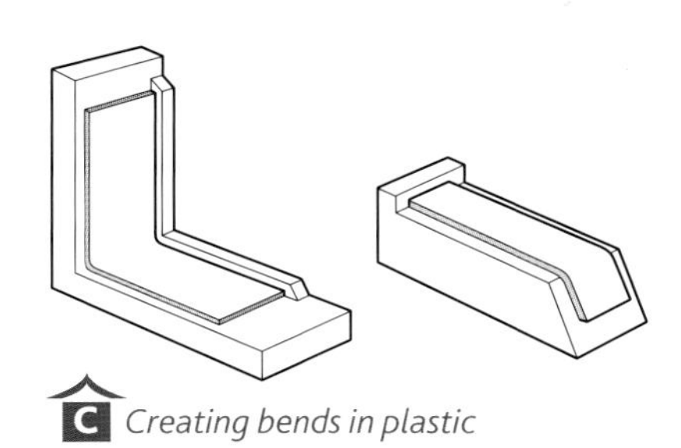

C Creating bends in plastic

Bending jig

Bending jigs can be used to make bends in a number of materials to improve accuracy and speed of operation. Like all jigs they should improve efficiency and ensure accuracy and cause no damage to the material being bent. The material that is being bent and its temperature will determine the material used to make bending jigs, that is, jigs for plastic can be made from wood while bending hot metals will require heat resistant material.

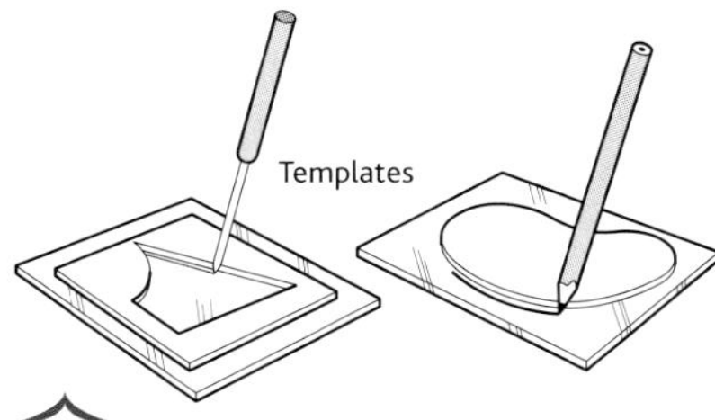

D Template for marking out a curved shape

Templates

These should be easy and simple to use and are made from paper, card and sheet materials. For repeated use they must be hard wearing enough to draw around, although paper templates can be glued to a material and then used to cut around and therefore are disposable. A template should save time from having repeatedly to measure and mark out.

Vacuum forming mould

Moulds can be made from a wide variety of materials (wood, card, clay) that are resistant to low level heat and provide the required level of surface finish. Complicated shapes can be formed but sides must taper slightly for ease of removal and any deep internal draws may need small vent holes. Vacuum moulds can be used repeatedly if manufactured from a robust material.

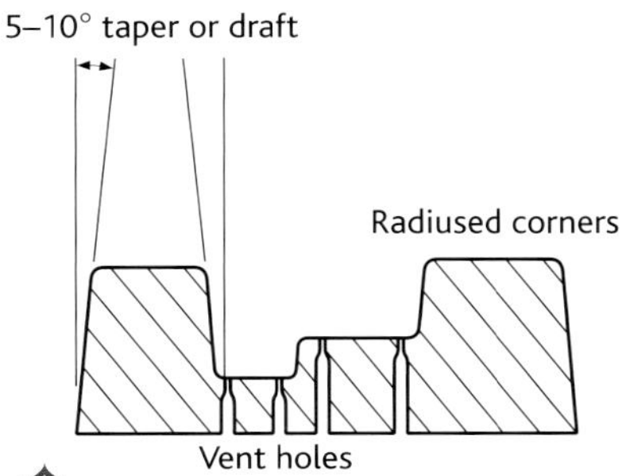

E Cross section of a mould

Activity

Develop a range of ideas that could be used to produce moulds for vacuum forming plastic separators for chocolates. Sketch a cross section of your vacuum forming mould.

links

Further information can be found on Moulding plastic, on page 126.

Summary

Jigs and templates are used in manufacturing to increase speed of operation.

Jigs and templates allow lower skilled workers to produce more accurate work.

At all levels of production manufacturers have to balance a number of factors to ensure efficiency.

AQA Examiner's tip

- You must make sure that you can explain how jigs/templates help with a range of manufacturing processes.
- You should understand and be able to explain levels of production and the relationship between the factors that influence the level of manufacturing efficiency.

12 Systems and control

12.1 Mechanical systems

Mechanical systems are a vital part of our everyday lives. The spoon that you ate your cereal with this morning is a very simple example of a **mechanical component**. The bus that brought you to school is an example of a very complex mechanical system.

Objectives

- Understand the four types of motion.
- Have knowledge of several mechanical components.
- Be able to able to link mechanical components together to form mechanical systems.

Levers

A lever is a bar that can be used to provide **mechanical advantage**. Levers are classified depending upon where the pivot point is placed. This also affects how the lever works.

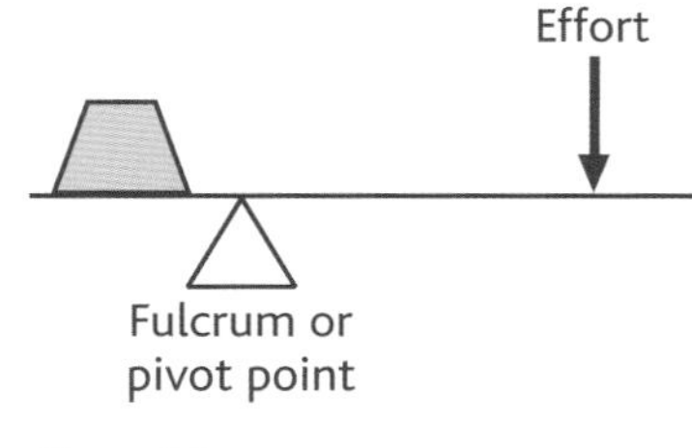

Class 1 lever

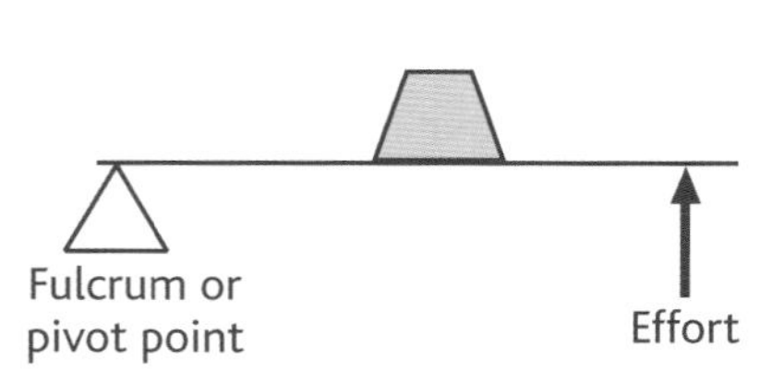

Class 2 lever

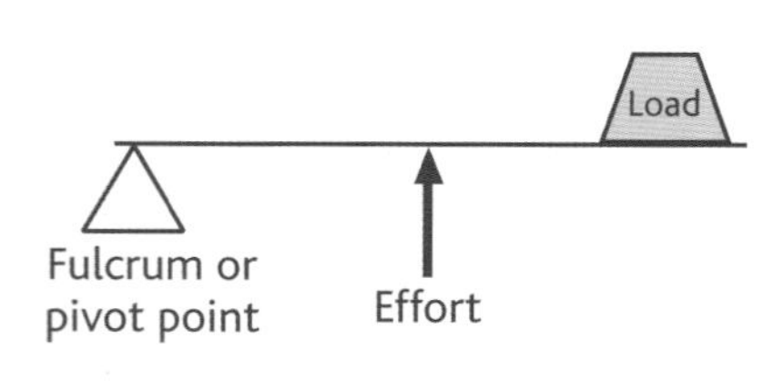

Class 3 lever

A *Types of lever: first, second and third class levers*

First class lever

Here the pivot point is placed between the effort and the load. The further the effort is placed from the pivot point the greater the mechanical advantage. The tree loppers are simply two first class levers joined together at the pivot point. They have a great mechanical advantage as the effort is a long way from the pivot point making it easy for the user to cut through branches.

Second class lever

Here the pivot point is placed at one end of the lever and the effort is at the other, leaving the load in the middle. The wheelbarrow is an example of a second class lever; again, the longer the handles are on the barrow the easier it would be to lift the load.

Third class lever

Here the pivot point is placed at one end of the lever and the load is at the other, leaving the effort in the middle. The spade is an example of a third class lever.

Key terms

Mechanical system: an assembly of mechanical components that form a machine.

Mechanical component: a mechanical part of a larger system or product.

Mechanical advantage: the way in which a machine makes things physically easier to do.

Linkages

A linkage consists of a number of levers connected together to form a mechanical system.

Reverse motion linkage

Pivot

Fixed pivot point

Pivot

Bell crank linkage

Fixed pivot point

B *Types of linkages: reverse motion and bell crank linkage*

Reverse motion linkage

When the top lever is pulled to the left the bottom lever is pushed to the right. The direction is reversed.

Bell crank linkage

When the bottom lever is pulled to the left the top lever moves down. The direction has been turned through 90°.

Types of motion

Before we begin to look at mechanical components and systems it is important for us to understand about the four types of motion.

Rotary motion

This is the easiest to understand. It is when a component is moving in a circle. Think about the wheel on your bike.

Linear motion

This is when a component is moving in a straight line. Think about a train on a railway track.

Reciprocating motion

This is when a component is moving backwards and forwards in a straight line. Think about the blade on a jigsaw.

Oscillating motion

This is when a component is moving backwards and forwards in an arc. Think about a swing or the pendulum on a grandfather clock.

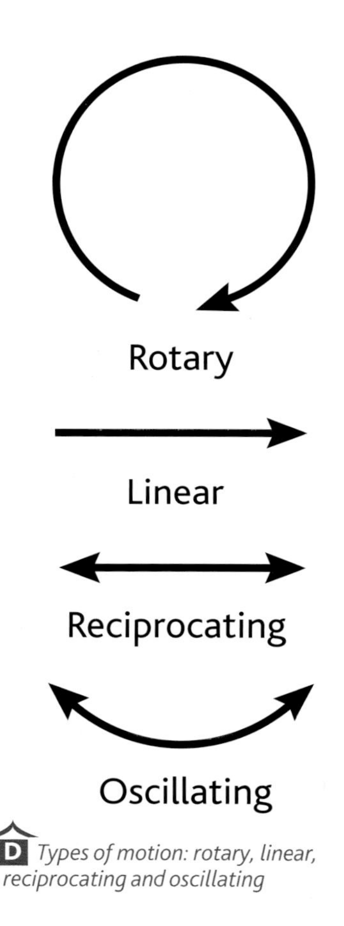

D *Types of motion: rotary, linear, reciprocating and oscillating*

CAMs

A cam is a shaped disc that rotates on a shaft. As it rotates it moves a follower up and down converting rotary motion into reciprocating motion. Different shaped cams affect the way that the follower moves.

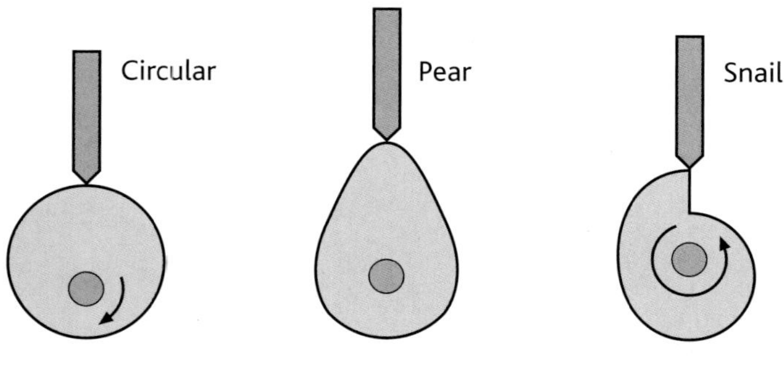

C *Types of cam: circular, pear and snail*

Circular cam

The circular cam makes the follower rise for 180° and then fall for 180°.

Pear shaped cam

The pear shaped cam makes the follower fall for 90°, is at rest for 180° and then rises for 90°.

Snail shaped cam

The snail shaped cam makes the follower rise for 360° and then snaps shut.

Transferring power and movement

Transferring power and movement is very common in many mechanical systems. Electric motors generate rotary motion and this movement needs to be controlled and transferred to where it is required. Think of a washing machine.

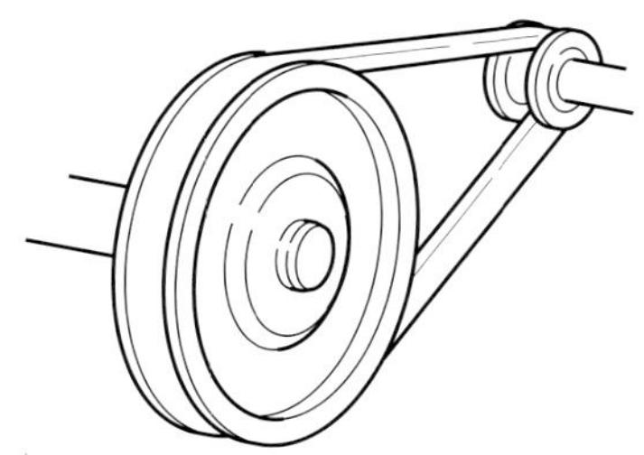

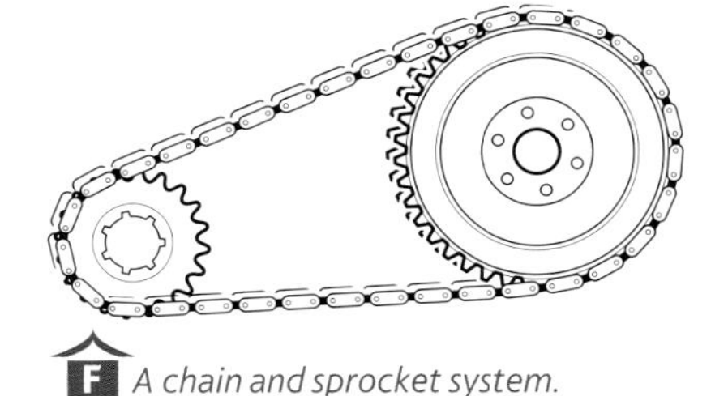

F A chain and sprocket system.

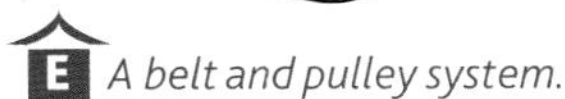

E A belt and pulley system.

Belt and pulley system

A belt and pulley system transfers rotary motion from one shaft to another. Your pillar drill will have a belt and pulley system. By using different sized pulleys we can alter the speed at which the shafts rotate. A small pulley driving a large pulley will decrease the speed. Belt and pulley systems are relatively inexpensive, lightweight and quiet mechanical systems.

Chain and sprocket system

A chain and sprocket system is another method of transferring rotary motion from one shaft to another. Your bicycle has a chain and sprocket system. By using different sized sprockets we can alter the speed at which the rear wheel rotates. Chain and sprocket systems are more expensive, heavier and require more maintenance than belt and pulley systems. However, they can transfer greater amounts of force and are less likely to slip or break.

Summary

There are three classes of lever: first, second and third class.

Linkages connect levers together to form mechanisms.

There are four types of motion: rotary, linear, oscillating and reciprocating.

Cams convert rotary to linear motion.

Belt and pulley, and chain and sprocket systems transfer rotary motion from one shaft to another.

Activity

1 Investigate the following machines. Make a list of the mechanical components and mechanical systems that they use.

a Mountain bike
b Pillar drill
c Tumble drier
d Scissors

AQA Examiner's tip

- You will need to be able to visually identify and name a range of mechanical components.
- You will need to describe how several mechanical systems work.
- You must understand the advantages and disadvantages of different mechanical systems.

12.2 Electrical systems

Electrical systems are a vital part of our everyday lives. From a simple hand-held torch to a CNC (computer numerically controlled) robotic welding arm, they all use electricity to power electrical components.

Power

The electricity you use will most likely come from a 9-volt battery. Mains electricity is too dangerous for use on school based prototype projects and should be reduced via a transformer.

Wires

The wires that connect your circuit are made from copper and covered in coloured Polyvinyl Chloride (PVC) for insulation and identification purposes.

Switches

A range of switches are available. The two most popular are the rocker switch and the push switch.

LED

LEDs (light emitting diodes) are a common form of low energy, safe lighting. They come in a range of colours and light intensities. If one LED is used with a 9-volt battery, the voltage needs to be reduced by using a resistor to prevent the LED from being damaged because they are only 1.5 volts.

Motors

A simple electric motor can add movement to your project when used in connection with a simple mechanical system.

Soldering

Soldering is a simple and effective method of joining electrical components together to form a circuit.

A *Soldering*

Objectives

Have a knowledge of several electrical components.

Be able to able to solder these components together to form a simple circuit.

Key terms

Electrical component: an electrical part that forms an electrical circuit, for example, batteries, switches and LEDs.

Electrical circuit: a number of electrical components connected together to form a functioning electrical product.

links

See Joining metals on page 118.

B *Electrical components*

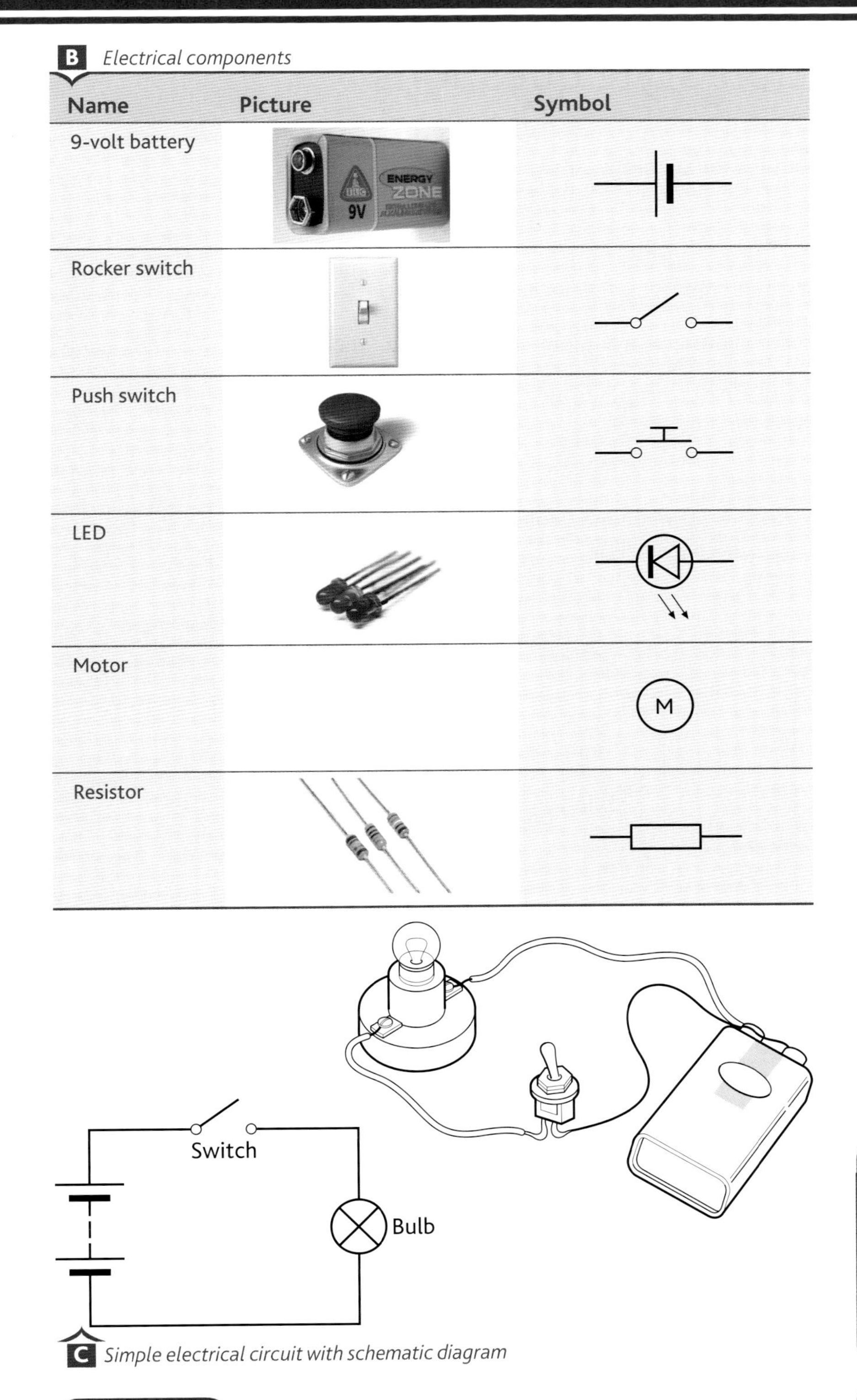

Name	Picture	Symbol
9-volt battery		
Rocker switch		
Push switch		
LED		
Motor		M
Resistor		

C *Simple electrical circuit with schematic diagram*

Activities

1. Make a single LED circuit complete with rocker switch, 9-volt battery and resistor.
2. Test your circuit.
3. Draw a schematic diagram of your circuit.

AQA Examiner's tip

- You will need to be able to name and identify a range of electrical components.
- You should be able to draw a simple electrical circuit.

Summary

Simple electrical components can be soldered together to form an electrical circuit.

AQA Examination-style questions

Test your understanding of processes and manufacture by answering the following exam-style questions.

Health and safety

1 Give **five** safety rules that you should observe when using a pillar drill. *(5 marks)*

2 Name **five** articles of PPE (personal protective equipment) and explain their use. *(10 marks)*

3 Explain the effects that the use of CAM (computer-aided machinery) has had on safety in the industrial workplace. *(8 marks)*

Techniques and process

4 The pictures below shows two designs for a notepad holder.
Both of them could be made in a school workshop.
Choose **one** of the notepad holders.

Notepad holder A

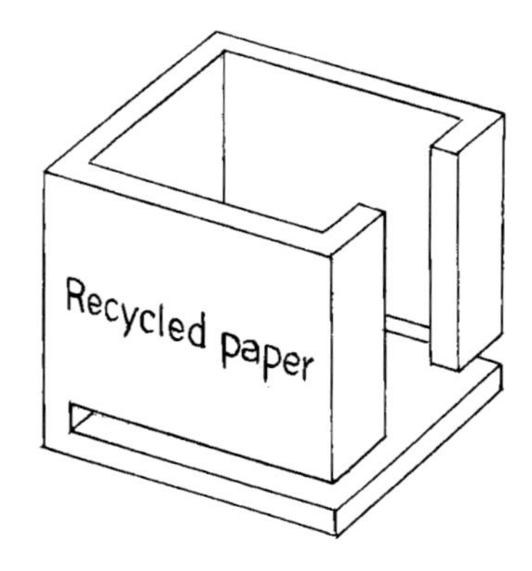

Notepad holder B

a) Name **one** suitable specific material you could use to make your chosen notepad holder. *(2 marks)*

b) Give **one** reason for your choice. *(1 mark)*

c) Use notes and sketches to clearly show how you would make a batch of **ten** notepad holders in a school workshop.

At each stage, name all the tools, equipment or software.

Stage 1: Marking out **or** CAD (computer-aided design) *(4 marks)*

Stage 2: Cutting and shaping **or** CAM (computer-aided manufacture) *(4 marks)*

Stage 3: Bending **or** joining *(4 marks)*

Stage 4: Finishing *(2 marks)*

Stage 5: Producing the 'recycled paper' text *(2 marks)*

5 Use notes and sketches to clearly show the process of vacuum forming. *(12 marks)*

6 Use notes and sketches to describe how the laminated frame of the chair, shown below, has been produced. *(12 marks)*

7 Clearly label a cross-section of a mould that is ready for aluminium casting. *(8 marks)*

Mechanical systems

8 Name the **four** types of motion and give an example of each. *(8 marks)*

9 Compare the use of a chain and sprocket system and a belt and pulley system. *(10 marks)*

10 Explain why it is important to maintain mechanical systems. *(6 marks)*

Electrical systems

11 Name **four** electrical components. *(4 marks)*

12 Produce a schematic diagram to show how the following electrical components would be connected to make a circuit. *(4 marks)*

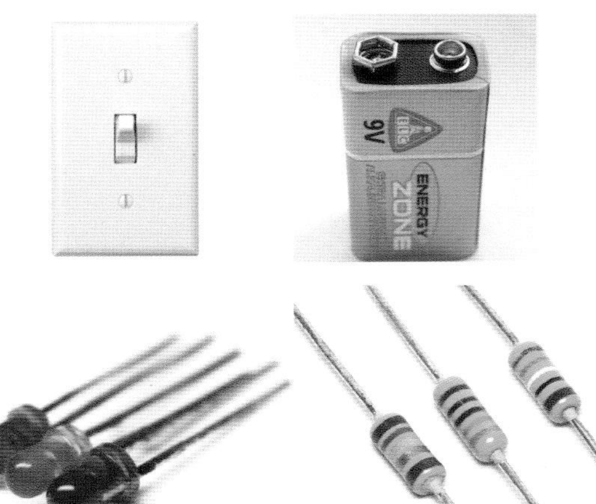

13 Explain why wires are covered in a coloured plastic. *(4 marks)*

UNIT TWO Design and making practice

The design and making practice unit is a very important part of your course. It is worth 60 per cent of your GCSE. You will be asked to design and make a product.

You will have approximately 45 hours to carry out this task. This may seem a long time, but you will need every minute of that time to:

- investigate and research the design context
- draw up a specification for the product
- create, develop, model and refine design ideas
- plan and make the product you have designed
- test and evaluate your product.

In this unit of this book you will find examples of students' work with commentary from the moderating team. The students' work will show different ways of tackling each part of the project. Moderators are employed by the exam board to check assessments of your work are in line with the national standards.

The design task

Your teacher will give you a design task or list of possible tasks. You will need to choose a task that interests you. You are likely to be working on this project for several months! You will need to use all the skills you have learnt whilst studying design and technology throughout your school life.

You need to show creativity in your designing. You may be able to improve on existing products. You may even be able to invent something completely new!

Your designs need to show consideration of environmental, social and moral issues.

Help with the controlled assessment task

You need to be familiar with the way that your project will be marked. This is called a scheme of assessment.

The scheme of assessment shows how many marks are available for each part of your work. It also describes the sort of things you need to include in each section.

Look at the different techniques in the students' work featured.

links

Look at the scheme of assessment in the official AQA document.
www.aqa.org.uk

The rules

You can only submit work which is yours! You can ask other people for opinions as part of your research, for example, asking people what they think of your design ideas, or what they think of the product you have designed. You cannot submit other people's design work. You need to show the source of any research work you include.

Your design work can be produced on paper or electronically.

If you use paper, A4 and A3 are both fine. The designer who designed the original Mini car sketched his ideas on a tablecloth! It is the quality of the designing that is important, not the size of the paper.

If you do your designing electronically, PowerPoint works really well. You could also create a single PDF file that contains your whole design folder. Either of these routes allows an examiner to look through your work easily. Make sure scanned sketches show up clearly. Remember to keep a back-up that you update every lesson.

Good luck!

If you work well, you should be able to produce design work and a product that you can be proud of. The work will show a very wide range of skills that could be used to impress a potential employer or a college interviewer, even where the job or course you are applying for is not related to design technology!

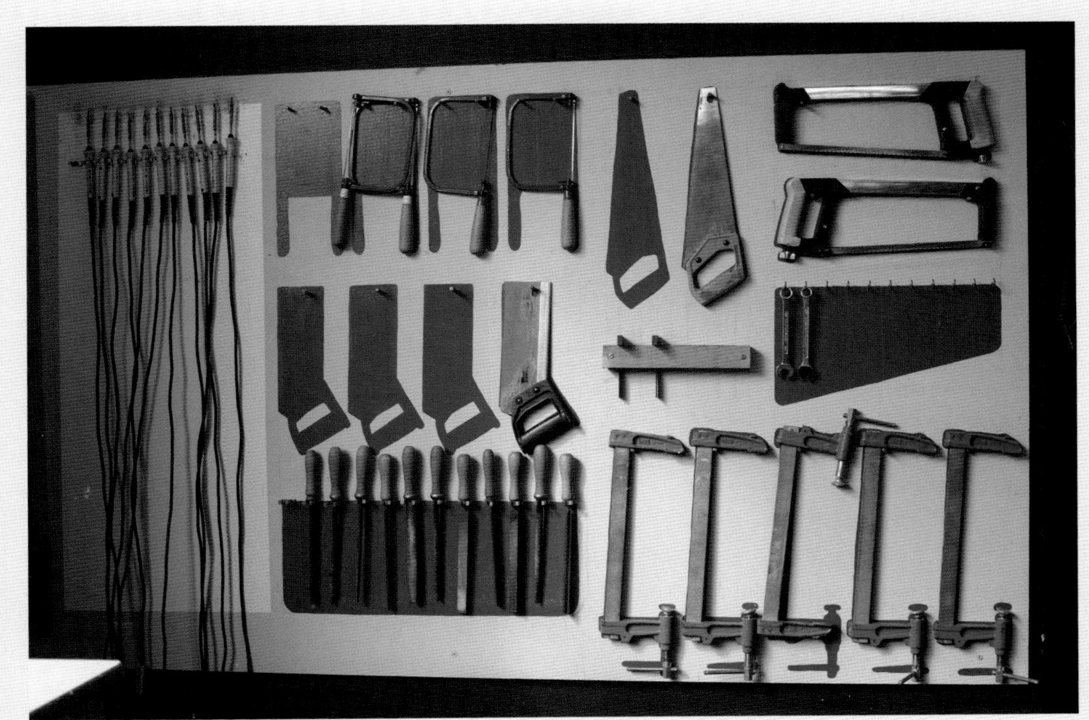

kerboodle!

13 Investigating the design task

13.1 Investigating the design task

Choosing the best design task is very important. You are going to be spending a large amount of time designing and making your project and it is worth 60 per cent of your final grade. Get it right and you will enjoy the experience, produce your best ever work and get excellent results. Your work on investigating the design task is worth up to 8 marks.

Objectives

- Understand several different methods of researching the design task.
- Understand how to analyse your research.
- Be able to produce a design criteria list.

Getting started

Your teacher will talk to you about the possible range of design contexts that are on offer to you. The list will be made up of design contexts that have been tried and tested to bring out the best results from GCSE students.

You should now analyse the design task that you have chosen. One of the best ways of doing this is by producing a thought shower. Place the design task in the middle of your page and think about all the information you will need to gather together before you can begin designing. Arrange this information around the central title. The thought shower should guide you into the research stage.

This student is designing and making a chess board for the teenage market. He has analysed the design opportunity and produced a broad range of areas for investigation. He has then refined the areas of investigation to more specific tasks.

AC5 communication

Good use has been made of Word to produce a clear diagram.

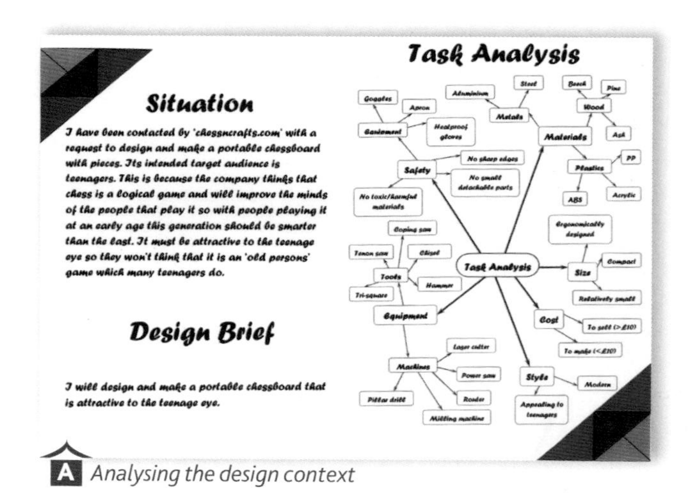

A *Analysing the design context*

Research

There are three things to stress about research; it must be relevant, analysed and concise. Pages and pages of unrelated research are worthless.

Market research

This is a very popular and worthwhile place to start your research. A designer will always look at existing products that are 'on the market' before they begin designing. The internet is a good research tool for this activity, but you can also use magazines, go and take pictures of products yourself or bring examples into school. It is ideal to have personal contact with the product you are analysing if possible – remember an Internet image gives only limited information. You must now analyse this research by commenting on each picture. ACCESS FM is a useful analysing tool.

links

See Product analysis on page 80 to remind yourself what ACCESS FM means.

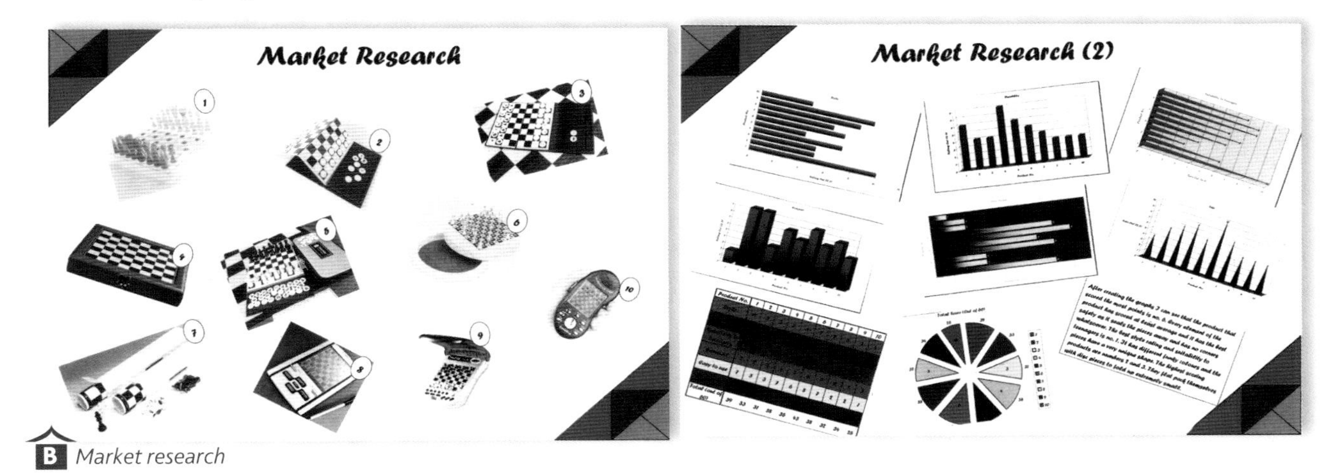

B *Market research*

AQA Examiner's comment

This student has researched the 'market' and found a wide and very diverse range of chess boards.
Each chess board has been carefully evaluated for its suitability. Initially this has been done using an evaluation grid and then by graphical representation.

AC5 communication

Good use has been made of the internet for researching chess boards.
Good use has been made of Excel to produce a spreadsheet and graphs that have then been imported into Word.

Survey/questionnaire

This is a useful way to find out what your target market wants from your product. You can produce a list of questions and either ask people in your target market to fill in a questionnaire or ask them directly. Record your results, analyse them and produce conclusions.

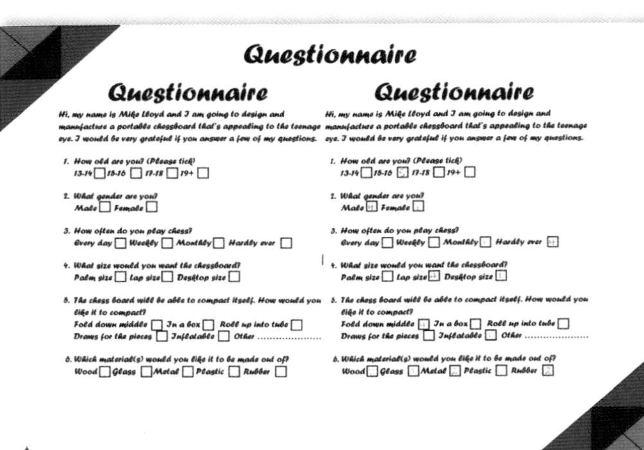

Questionnaire

Questionnaire

Hi, my name is Mike Lloyd and I am going to design and manufacture a portable chessboard that's appealing to the teenage eye. I would be very grateful if you answer a few of my questions.

1. How old are you? (Please tick)
13-14 ☐ 15-16 ☐ 17-18 ☐ 19+ ☐
2. What gender are you?
Male ☐ Female ☐
3. How often do you play chess?
Every day ☐ Weekly ☐ Monthly ☐ Hardly ever ☐
4. What size would you want the chessboard?
Palm size ☐ Lap size ☐ Desktop size ☐
5. The chess board will be able to compact itself. How would you like it to compact?
Fold down middle ☐ In a box ☐ Roll up into tube ☐
Draws for the pieces ☐ Inflatable ☐ Other ....
6. Which material(s) would you like it to be made out of?
Wood ☐ Glass ☐ Metal ☐ Plastic ☐ Rubber ☐

Questionnaire

Hi, my name is Mike Lloyd and I am going to design and manufacture a portable chessboard that's appealing to the teenage eye. I would be very grateful if you answer a few of my questions.

1. How old are you? (Please tick)
13-14 ☐ 15-16 ☐ 17-18 ☐ 19+ ☐
2. What gender are you?
Male ☐ Female ☐
3. How often do you play chess?
Every day ☐ Weekly ☐ Monthly ☐ Hardly ever ☐
4. What size would you want the chessboard?
Palm size ☐ Lap size ☐ Desktop size ☐
5. The chess board will be able to compact itself. How would you like it to compact?
Fold down middle ☐ In a box ☐ Roll up into tube ☐
Draws for the pieces ☐ Inflatable ☐ Other ....
6. Which material(s) would you like it to be made out of?
Wood ☐ Glass ☐ Metal ☐ Plastic ☐ Rubber ☐

C *Questionnaire*

AQA Examiner's comment

This student has produced a range of questions that will help him understand the requirements of his target audience. He could have extended his range of questions to get more detailed information.

AC5 communication

Good use has been made of Word to produce a questionnaire.

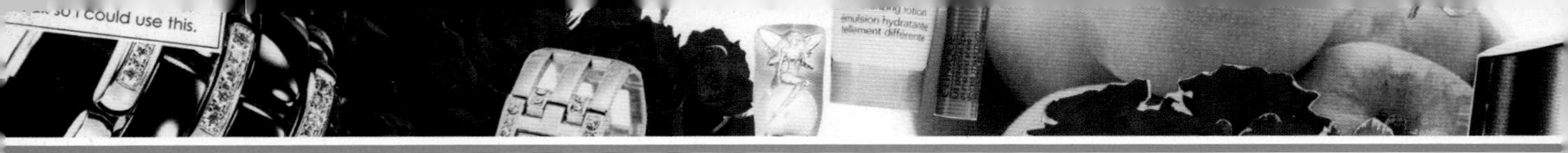

Anthropometric data

The chances are that people are going to have to interact with your design when it is in use. Therefore you need to consider anthropometric data and use this information to make sure that your design is ergonomically correct.

links

See Form follows function on page 52 to remind yourself about anthropometrics.

Mood board

A mood board is a useful way for you to get an understanding of the style, colour and shape that suits your target market. This can be done by arranging images of the sort of things that interest your client. Again, it is important to analyse this information so you can then use it to influence your designs.

D *Mood board*

This student is designing and making a jewel box. In consultation with the client, she has produced a mood board that shows the likes, interest and styles that appeal to the client.

AC5 communication

Good use has been made of the internet as a research tool.

Good use has been made of Word to present the information.

Sizes and dimensions

It may be useful for you to find out some relevant sizes and dimensions for your project. For example, if you are making a CD rack it will be essential to find out the exact sizes of a CD. Record these in your design folder.

Visits/interviews

It may be possible for you to go and visit your target market, for example, you may visit a primary school if you were making a child's educational toy. It may be possible for you to photograph the visit, but you must ask permission.

Design criteria

The design criteria are a list of the functional and aesthetical requirements of your design. It will give you a focus for you to begin designing and enable you to carry out further research by modelling, investigation into materials, finishes and methods of manufacture.

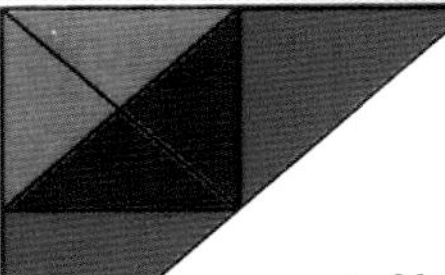

Specifications

- **Must be attractive to the teenage eye**
  It is specifically being designed for teenagers so it must be eye-catching to this certain generation.
- **Must be a suitable size**
  It must be small enough to take around without it being a nuisance but must be big enough so it isn't hard to play.
- **It must be cost effective**
  It should be of good quality but it must be cheap enough to make, sell and get a profit out of it.
- **Must have no sharp edges**
  This is a safety precaution; it must not have sharp edges to minimize potential injuries.
- **Must be secure**
  Even though it is portable, it still must be secure and stable. It needs to be secure enough to be able to play without the thought in the back of the mind that a slight knock can ruin the whole game.
- **Must be ergonomically designed using anthropometric data**
  It must be designed so teenagers will be able to grasp the pieces with ease without knocking other pieces on the board.
- **Must be easy to use**
  It must be easy to get out of its compacted state, easy to play and then easy to put away. Also the quicker it can be taken out and put away the better.
- **Must be well made/durable**
  It must of good enough quality to last quite a long time (about 7 years as this how long the teenage era lasts) without falling apart or breaking.
- **Must be environmentally friendly**
  This includes making and selling. It must not be made using harmful materials/chemicals e.g. asbestos.
- **Must be stylish**
  Teenagers will only buy this product if they think it is stylish. It could be themed or attractively designed.
- **If it is being made out of wood, it must not splinter**
  If the product splinters it will not only not be very durable but it could also cause harm to people who into contact with it.

E *Design criteria*

AQA Examiner's comment

This student has identified a number of design criteria. Each point is clearly explained and has been generated from the research that was previously undertaken.

Summary

Choosing the correct project is very important.

There are a number of types of research that can be undertaken. This could include: market research, questionnaire, survey, anthropometrical data, mood board, sizes/dimensions, visits/ interviews.

Design criteria will help you begin designing.

14 Development of design proposals

14.1 Development of design proposals

Developing and modelling your design proposals is a major part of your controlled assessment task. This section is worth up to 32 marks from a total of 90 marks for the whole project. You need to be able to clearly communicate your design ideas with other people. To do this you can use a range of techniques including sketching, model making and computer-aided design. The next 10 pages show examples of pupils' work demonstrating a range of techniques for developing design proposals.

It is important to explain what you are doing and why, because the examination board will look for a planned approach to your designing. You also need to consider social, moral, environmental and sustainability issues in your designing.

Creativity

Creativity is an important part of work in resistant materials technology. If you show innovation, originality and creativity in your design, it will allow you access to higher marks for your designing. You can use a range of techniques like the ones here. These techniques may allow you to come up with new ideas that you may not have thought of otherwise.

Objectives

- Create and communicate a range of imaginative and innovative ideas.
- Consider social, moral, environmental and sustainability issues in designing.
- Develop ideas through experimentation using a variety of techniques and modelling.
- Use CAD as appropriate.
- Select appropriate materials and components.
- Specify full details of the product and its manufacture.

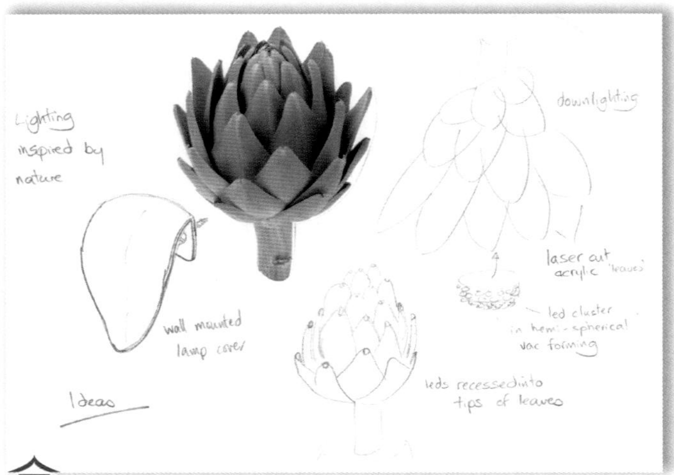

A *Inspiration from nature*

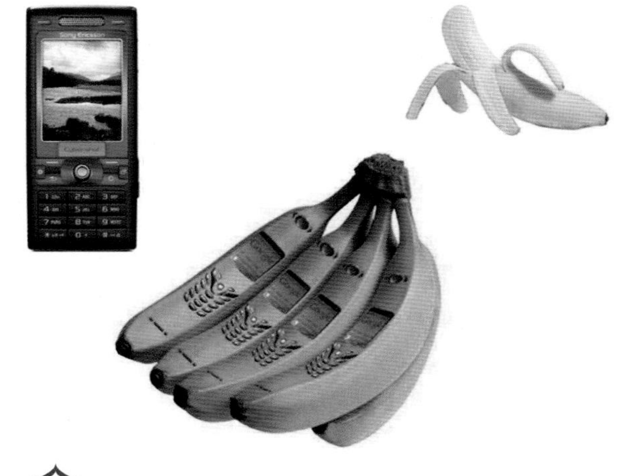

B *Using morphing*

This candidate has used an artichoke for inspiration for a lampshade.

Two very different products have been combined to create an original design.

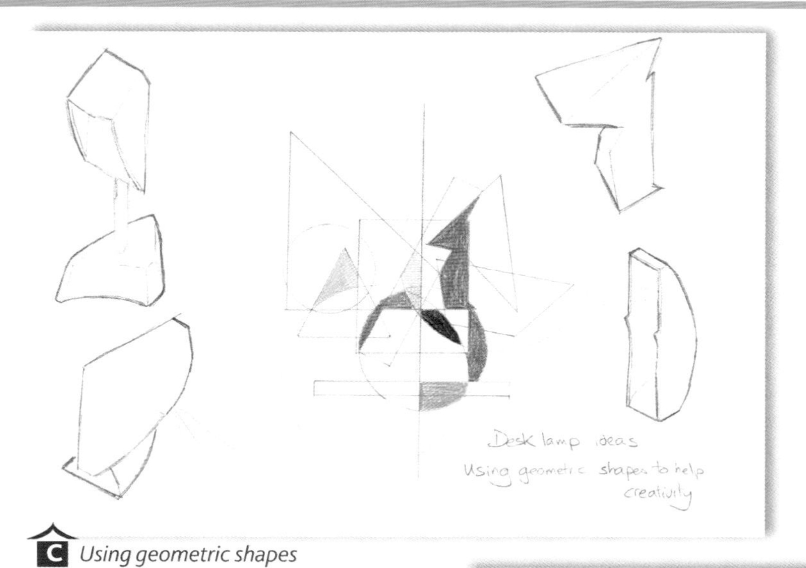

C *Using geometric shapes*

links

See more information on social and moral issues on page 68 and sustainability and environmental issues on pages 70–71.

Scruffiti, jackstraws and geometrical shapes have been used for inspiration for desk lamp designs. Without using these techniques, it is likely that the design ideas would have been much less creative.

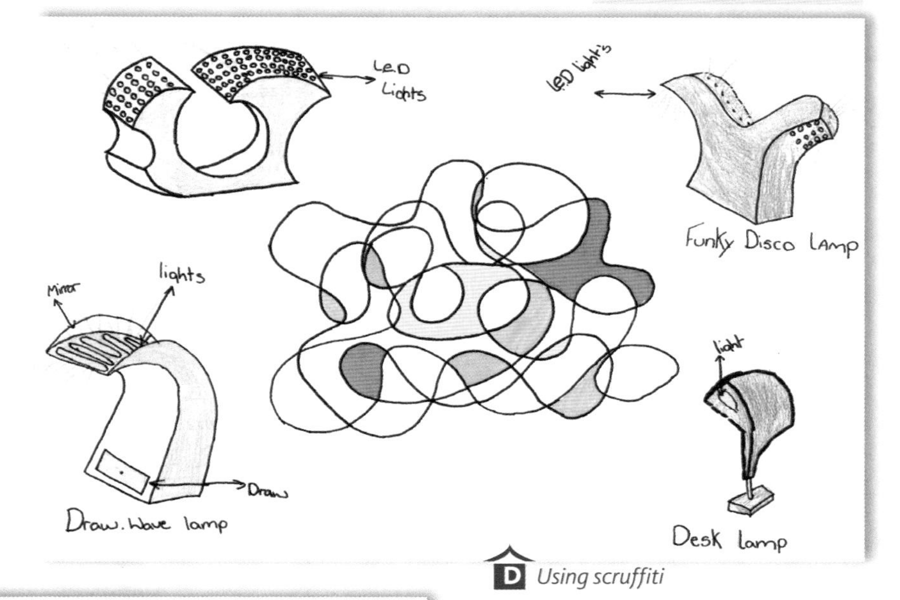

D *Using scruffiti*

AC5 communication

Use clear, concise labels on your design sketches to explain what is not obvious from the sketch.

links

See more information on these techniques on page 62.

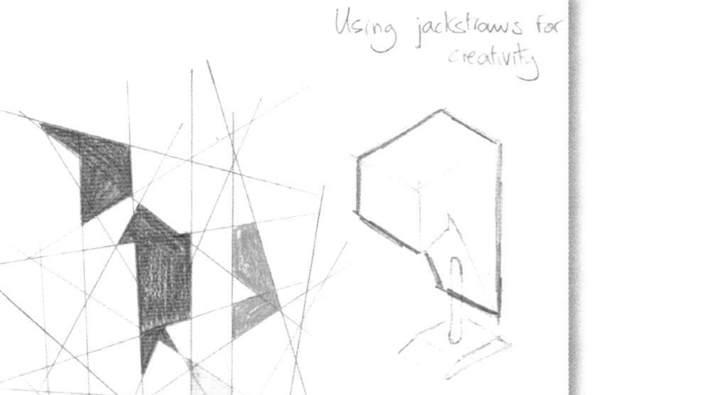

E *Using jackstraws*

14.2 Using 3D sketches and cross sections

On the previous two pages, there were examples of different techniques to achieve creativity. Here, initial design proposals are being developed. This means that the pupils are starting to add detail to their designs. They are starting to think about what materials to use for different parts, and how these parts can be joined. The shapes of the designs are being modified and improved. It is important to add brief notes to explain why changes are being made.

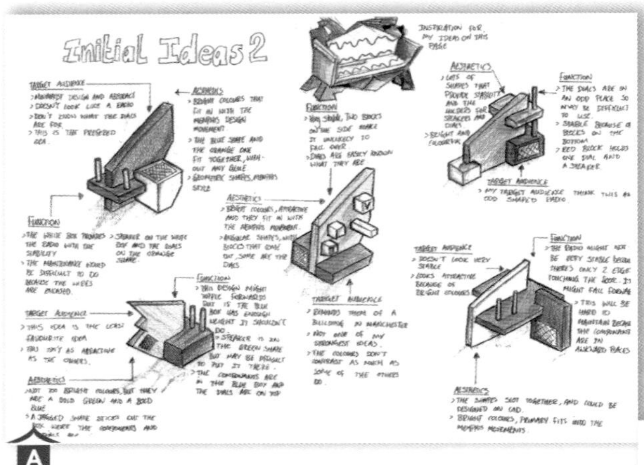

A

Examiner's comment

This is a second page of ideas from a folder. It shows the candidate beginning to focus ideas and analyse them.

AC5 communication

Notes on design and development pages need to be concise, relevant and legible.

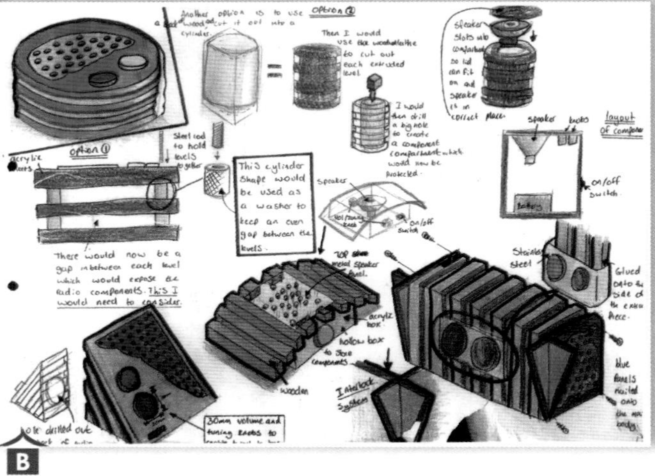

B

Examiner's comment

This candidate has produced a variety of different ideas. They have considered the construction of each design.

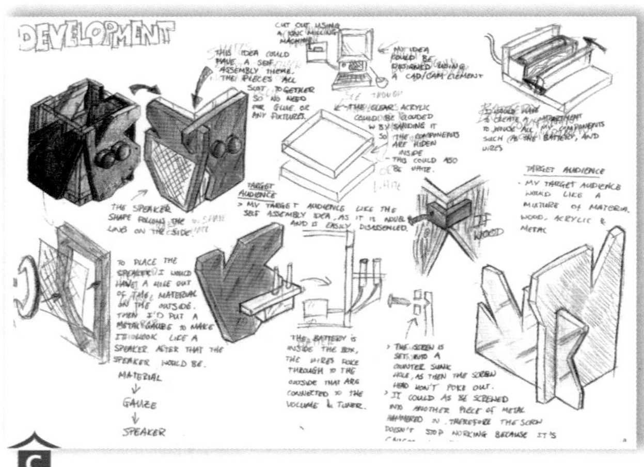

C

Examiner's comment

Alternative construction methods need to be considered during the development stage.

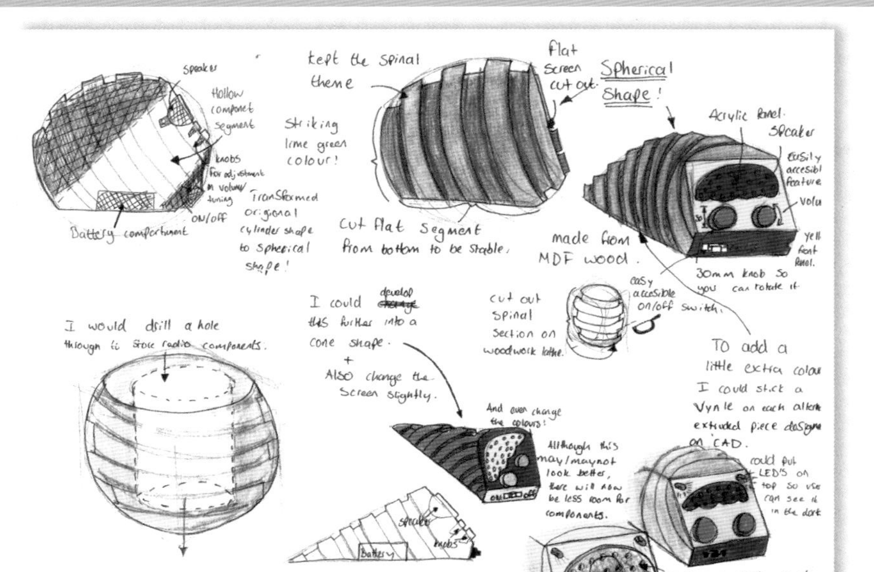

On this page, slight modifications are being considered for a chosen design.

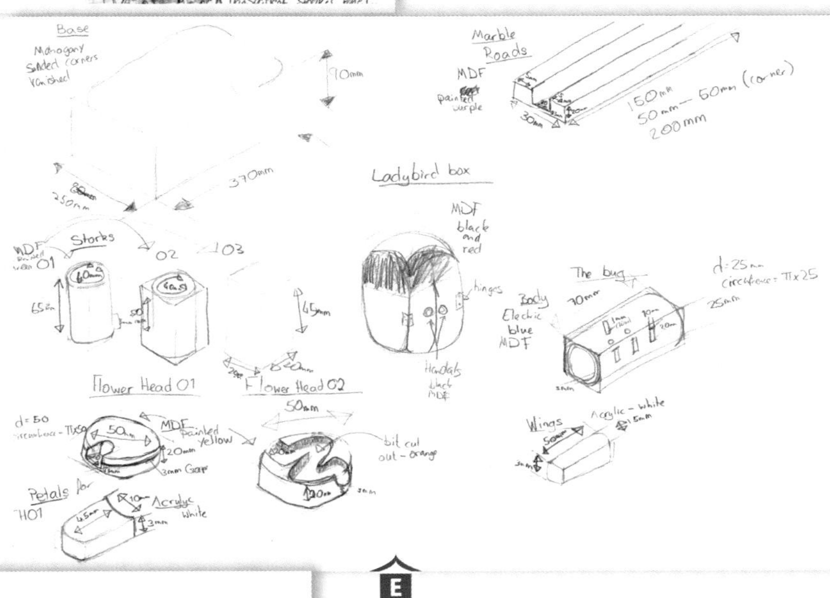

This candidate is considering the dimensions of each part of their design.

E

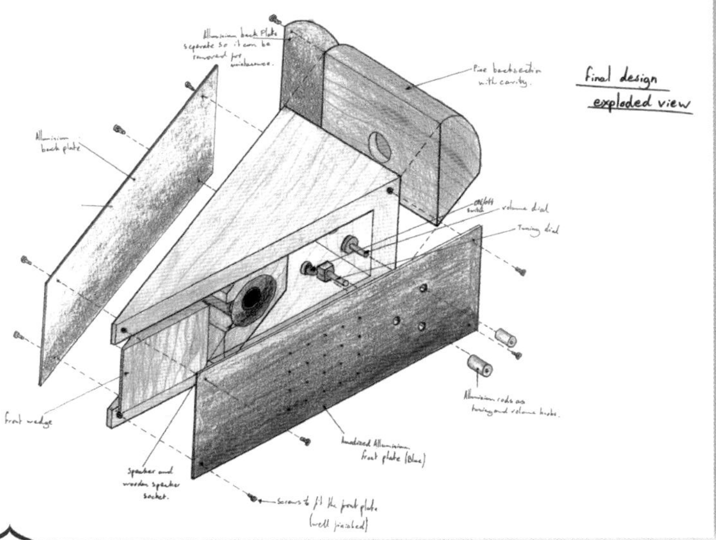

An exploded sketch is excellent for showing how parts and components will be fitted together.

14.3 Modelling

Making a series of models is an excellent way of refining and improving your ideas. The pages from design folders here show different ways of modelling. Each candidate has shown through sketches or notes how they have refined their ideas.

Examiner's comment

Models have been used here to demonstrate a range of design ideas. The designs were illustrated in PowerPoint and presented to potential consumers.

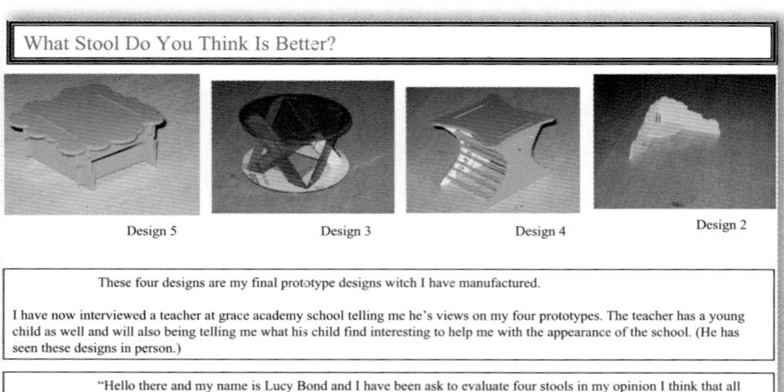

What Stool Do You Think Is Better?

Design 5 | Design 3 | Design 4 | Design 2

These four designs are my final prototype designs witch I have manufactured.

I have now interviewed a teacher at grace academy school telling me he's views on my four prototypes. The teacher has a young child as well and will also being telling me what his child find interesting to help me with the appearance of the school. (He has seen these designs in person.)

"Hello there and my name is Lucy Bond and I have been ask to evaluate four stools in my opinion I think that all four stools are all genuine designs, however there is one which catches my eye. That is design 4. What I like about this design is that it is an appearing and busy stool to look at. If this stool was filled with bright colors' then it would make the stool busier at which young children like as I know from my aged 4 son. My son loves bright colors' and funny shaped objects which design four is. It is a funny, extra ordinary shape. My son also likes running his hand and he's fingers up bumpy objects, design four has those supporting bars witch can be used as a toy for my son. You asked me what my son found interesting. Simply cartoons. My son loves all the cartoons and I am sure that one of my son's favorite cartoons would go well on one of these stools will surely keep him happy."

A

AC5 communication

Sketches and models communicate ideas without the need for lots of writing. This helps to keep the folder concise.

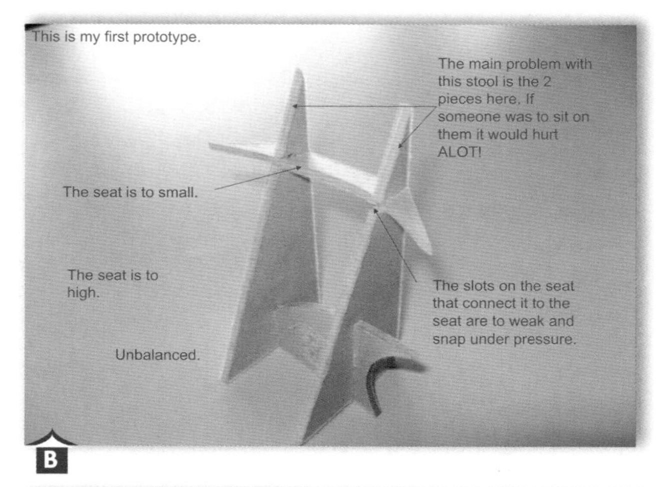

B

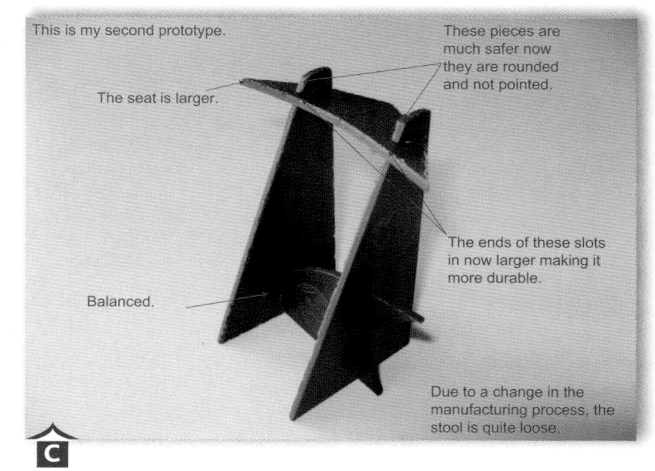

C

D

Examiner's comment

Three models made by the same candidate. They have been used to improve and develop the original idea. Notes have been added to explain changes made to each model.

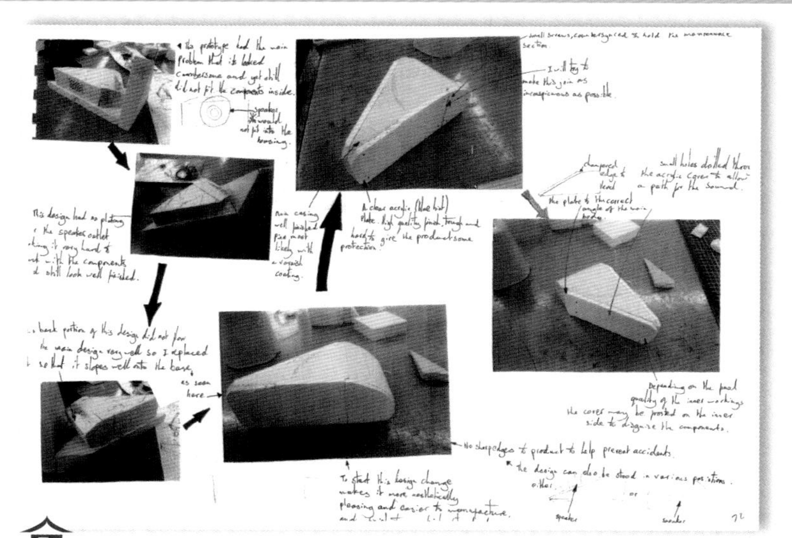

E

This storyboard shows the development of a design idea. Notes explain how the ideas have been refined at each stage. The candidate has used mainly styrofoam with some foamboard.

Model made from card, dowel and clear self-adhesive tape. The model has inspired the candidate to explore alternative construction methods and sketch them around the original model.

F

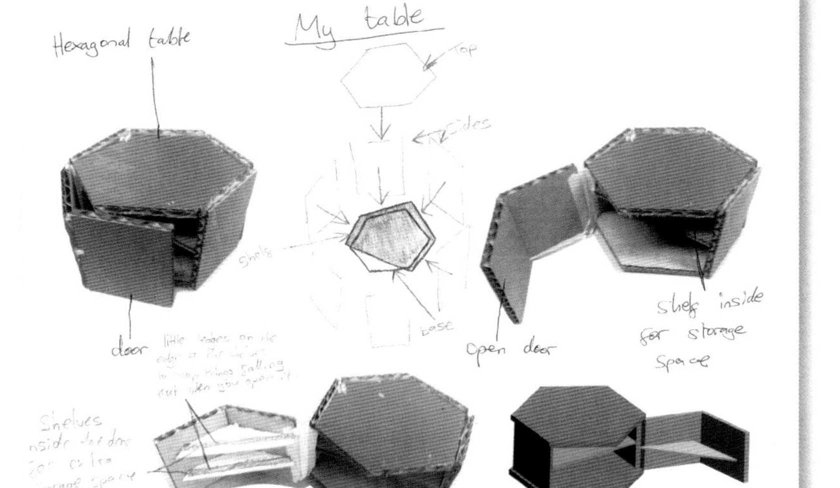

This cardboard model was cut using a laser cutter. The candidate developed the design by adding sketches onto and around the photograph of the model. (See shelves in door.) The design has been further developed in the Pro/DESKTOP design.

14.4 Computer-aided design (CAD)

Computer-aided design (CAD) offers an excellent way to see and refine design ideas, without having to cut and join materials. Some software allows moving parts to be demonstrated. Three-dimensional CAD software allows a design to be rotated and viewed from different angles. Different surface finishes can be compared, both by you and by your target market. Final designs can be printed on paper with great accuracy. CAD/CAM equipment can make the parts quickly and precisely.

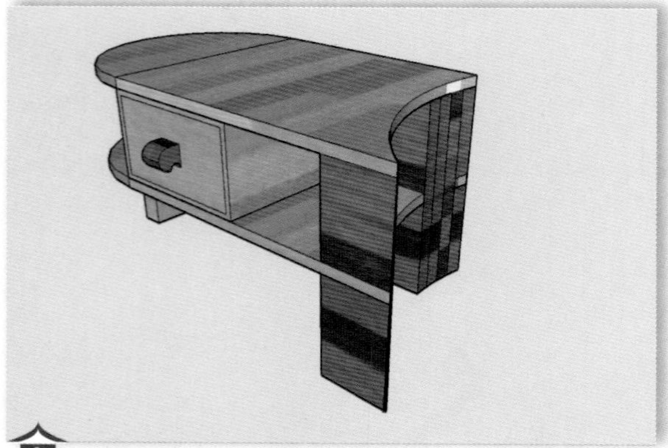

A Coffee table in Google SketchUp

links

Look at the Kerboodle website for further examples of CAD.

Google SketchUp works well for illustrating designs.

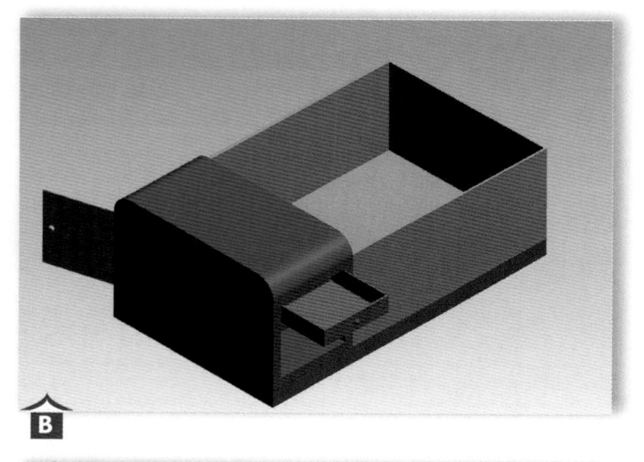

B

C

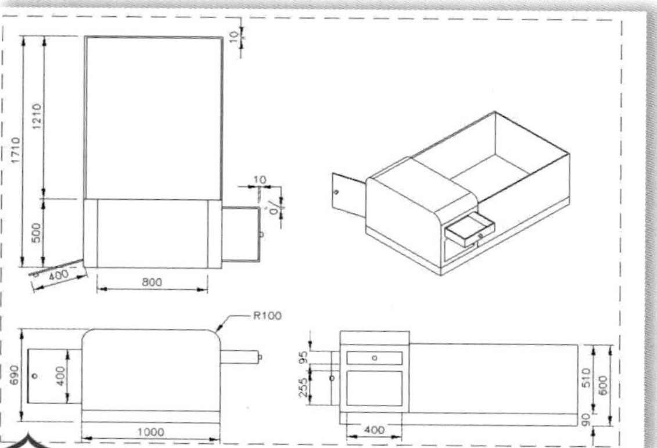

D A combined coffee table/fish tank has been modelled in Pro/DESKTOP design

The Pro/DESKTOP photo album allows rapid viewing of different surface finishes. A Pro/DESKTOP engineering drawing can be created from the design in seconds. In a few minutes more, dimensions can be added.

AC5 communication

Being able to see 3D models on-screen avoids the need for extended written descriptions.

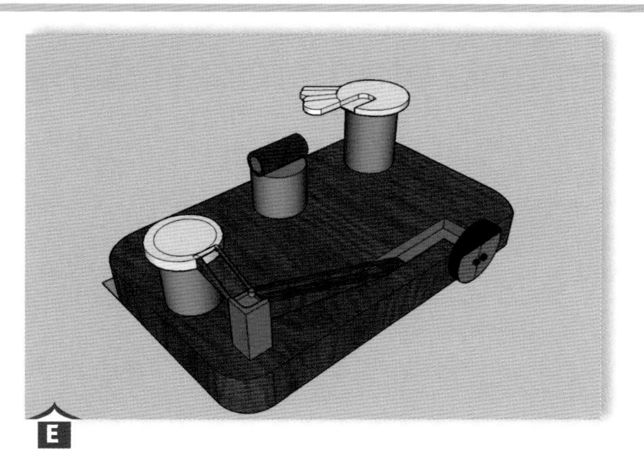
E

F

G Model and CAD versions of marble run

From a foamboard model, this candidate used Google SketchUp for an initial virtual model (Source **E**). A moving part has been developed using Pro/DESKTOP design (above).

It is important to record evidence of any development work done using computer-aided design. This candidate has pasted each modification into Word and added notes to explain improvements.

Developments I have made to my design whilst drawing it in CAD.

Originally I put the rails for the sliding mechanism all the way along the bottom to the tabletops. I then made a cut out from these rails for the table legs – I felt it would look better if the table legs were on the bottom of the table tops, it would also make the table utilise more space (for the first time buyer, my target buyers), and also make the table more secure when assembled.

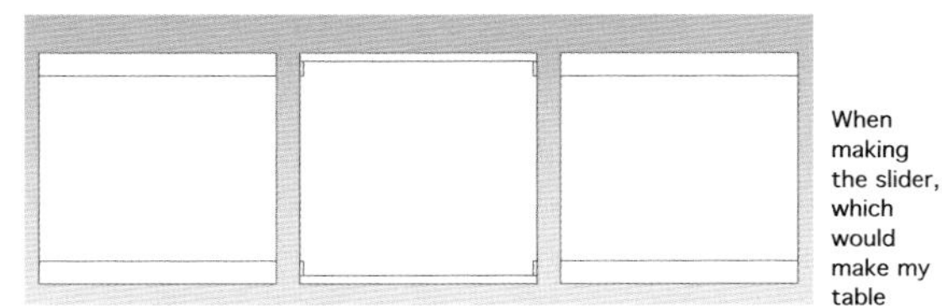

When making the slider, which would make my table extendable, I changed the length of it. This is because the shorter mechanism would be out of proportion with my table – one feature of my table design is to have gaps between the tabletops, with a shorter rail this would not be possible. I also lengthened the slider, as the shorter slider would not be able to hold the tabletops.

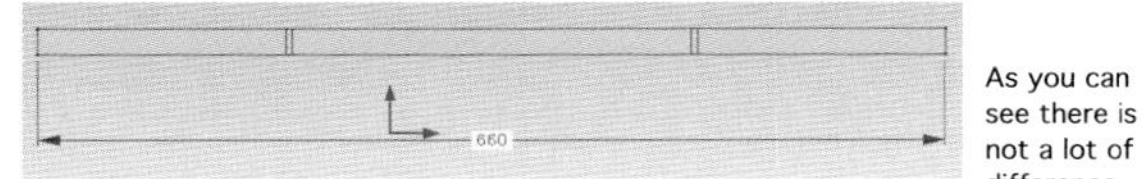

As you can see there is not a lot of difference in the length of the slider but there is in the distance of where the slots are to hold the middle tabletop in place. The more of the slider holding up the end table tops the more balanced and secure the table will be.

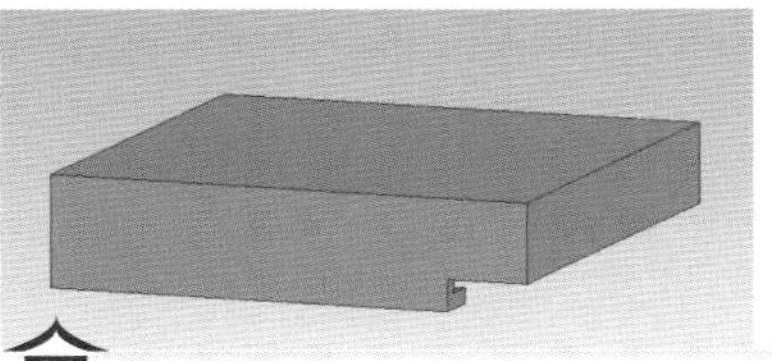

When I first drew the tabletops I drew them all as one sketch. I edited the sketch as when I assemble my parts and show the sliding mechanism of my table (when the table is smaller) it was not possible to solely remove the middle table top by itself.

What I did to resolve the problem.

I changed the length of the slider, which then in turn meant I had to change where the rebate would be in the end table tops.

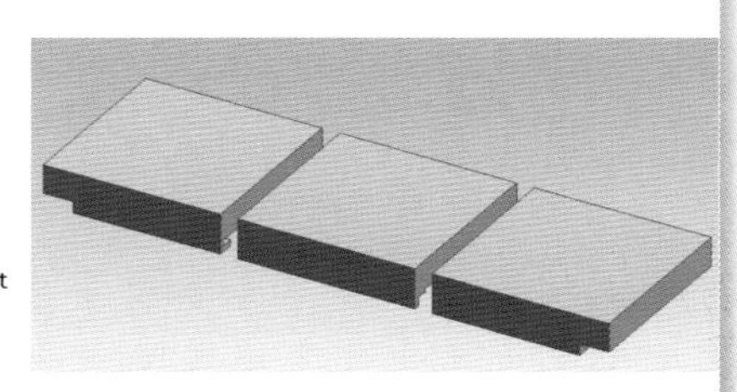

(The rebate is for when the table is full length to stop the slider moving around.)

When I went to assemble my table in the smaller form I found that the slider mechanism was too long. I aligned the slider with each end of table but there was a massive gap between the tabletops.

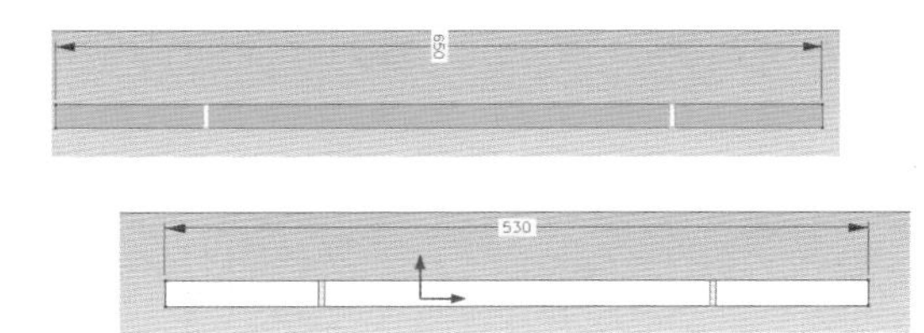

H

14.5 Presenting formal ideas and plans for making

You should present your final design so that there is enough information to allow somebody else to make the product (even though you are going to make it yourself!). You should include an accurate drawing, a list of materials, sizes, finishes and components needed, and a plan for making.

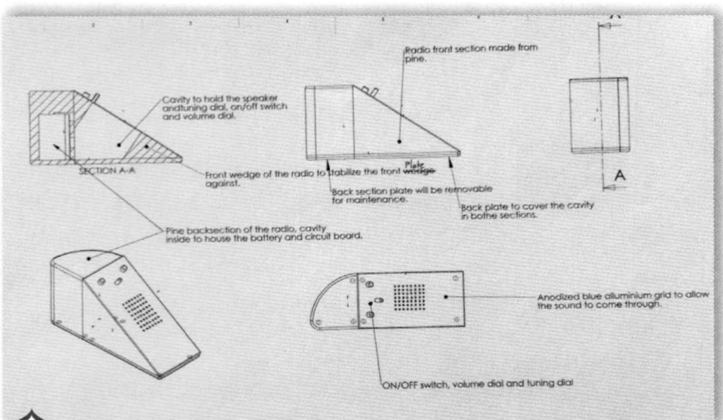

A *The formal drawing includes notes with technical information, but no sizes have been included*

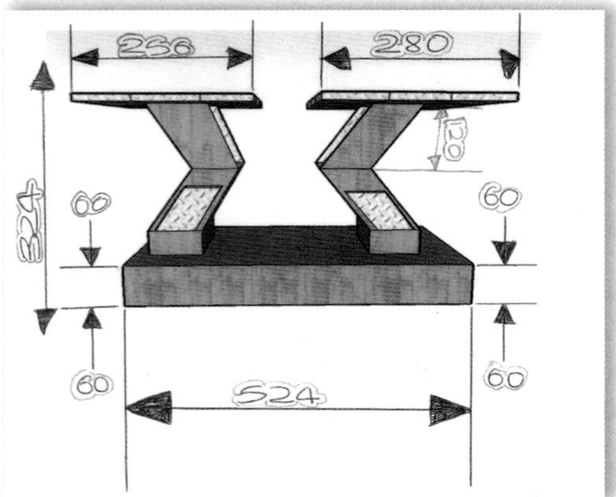

B *Sizes have been added by hand to the Google SketchUp print-out above*

C

links

Look at the online exercise on planning for making and using a flowchart.

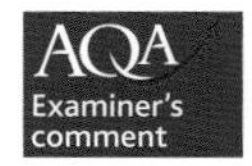

A cross section is a useful way of showing constructional detail.

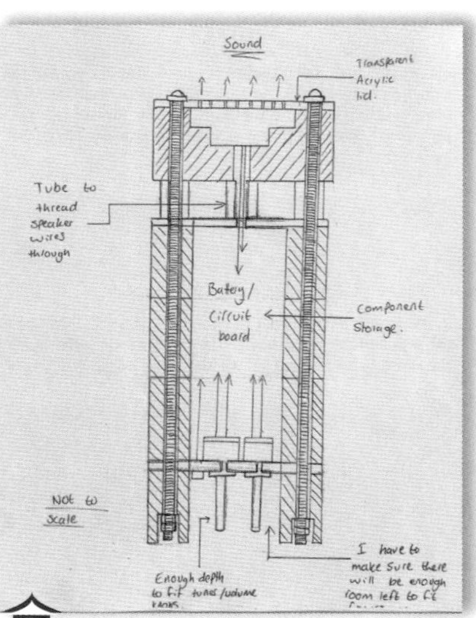

D *Cross section sketch*

Part	No off	L	W	Th	Material	Finish
Legs	4	400	25	25	steel tube	spray Hammerite
Rails	2	450	25	25	steel tube	spray Hammerite
Strut	1	500	25	25	steel tube	spray Hammerite
table top	1	550	500	18	mdf	emulsion
Screws	12	40	4		steel	

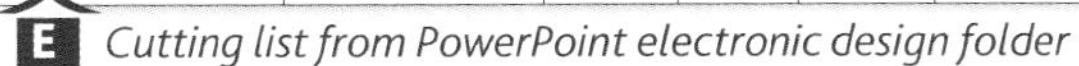
E *Cutting list from PowerPoint electronic design folder*

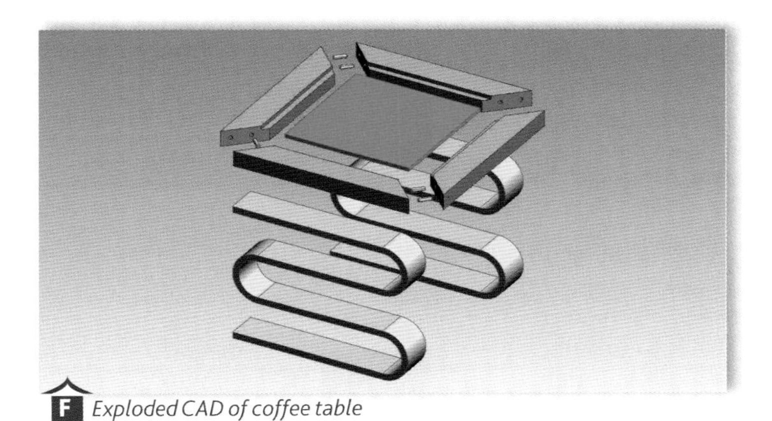
F *Exploded CAD of coffee table*

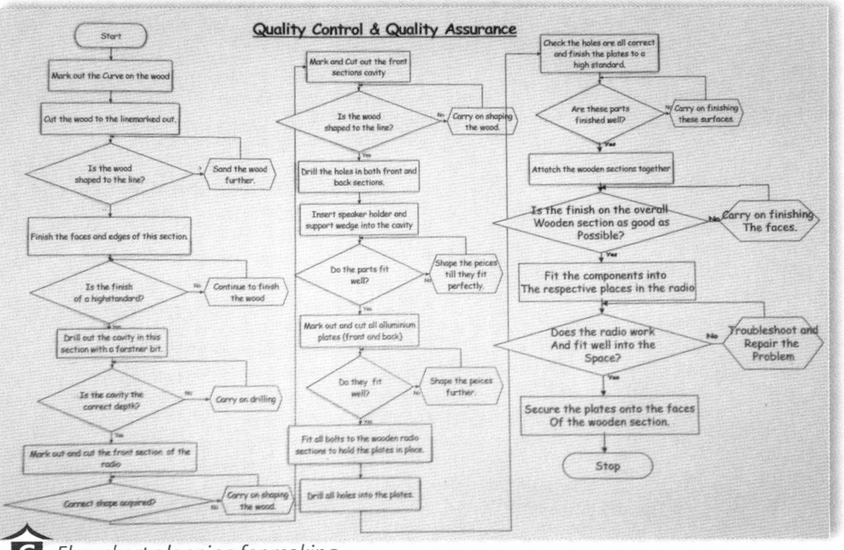

G *Flowchart planning for making*

A cutting list saves time and resources as you start to make.

Pro/DESKTOP designs assembled from separately drawn parts are easy to 'explode' to show detail.

A flowchart is a very concise and clear way to communicate a plan for making.

AC5 communication

A cutting list and a formal drawing can be included on the same page of your folder. This helps to keep your folder concise.

Summary

Design ideas need to demonstrate creativity, flair and originality.

Designing needs to be carried out in a logical way.

Social, moral, environmental and sustainability issues need to be considered.

Ideas need to be developed through experimentation including using modelling and CAD.

Material choices need to be explained.

Accurate drawings, a cutting list and a plan for making are needed.

15 Making

15.1 Wood-based products

Getting started

You have already spent a great deal of time developing and modelling your design proposals. The making section is also worth up to 32 marks from a total of 90 marks for the whole project. This section can be both interesting and rewarding as you successfully plan and manipulate materials to produce your own design. To succeed in the making section you must try demonstrating your skills by using as full a range of techniques and processes as possible. It is very important that your work is completed by the deadline set by your teacher and to the highest level of accuracy you can possibly achieve.

Wood is an excellent and traditional material to use for manufacturing your ideas. Many of the tools required are readily available, and with care and practice accurate, high quality outcomes can be produced. The completed projects here show different approaches to making with each candidate producing an individual response to a design brief.

Develop and complete a solution to a problem that is predominately manufactured from wood.

The examples on these pages are made predominently from wood.

- Ensure you finish your final outcome and that you plan your time to ensure any necessary finishing technique is completed.
- The candidate in Source **A** has used a range of skills and tools to produce a traditional product but has shown evidence of a very high level of making and finishing.

A

- Try to make your project as demanding as possible by perhaps taking an unusual approach to an undemanding problem.
- Source **A** shows a candidate who has taken a relatively simple project and tried to produce a project with demand and complexity. Following testing a variety of jigs and templates have been produced to allow the manufacture of the bird box in quantities that can be packed and sold.

B

- Finish your final outcome to a high standard.
- Take care at each stage of making.
- The candidate in Source **B** has produced a quality outcome using a manufacturing technique that requires a high degree of skill to achieve consistent levels of accuracy.
- Each component was checked for accuracy before assembly.

C

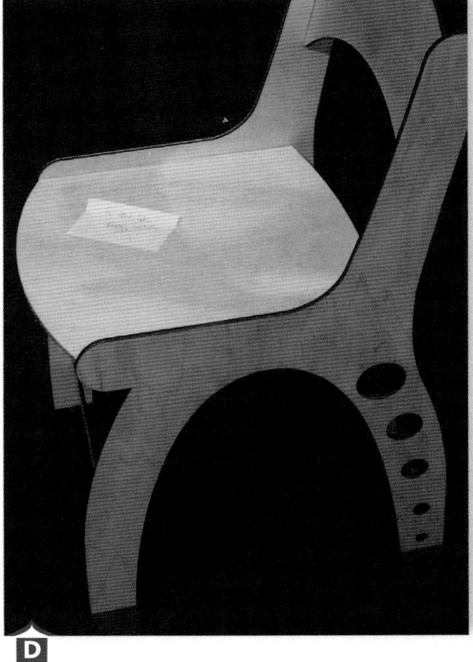

D

- In both Sources **C** and **D** candidates have tackled a similar problem but have produced items via very different manufacturing paths.
- Source **D** has been produced using only CAM but is carefully hand finished to ensure a quality level of quality. Source **C** uses more traditional skills but tackles a seating problem in an unusual manner by not using a typical under-frame.

Summary

Simple problems can be tackled in an innovative way to increase complexity in making.

It is important to produce a completed project that shows a high level of accuracy and a necessary applied finish.

Utilising CAM, where appropriate, can help achieve a higher level of accuracy and may allow access to higher grades.

15.2 Metal-based products

There are many different types and sizes of metals available to use for manufacturing your ideas. The tools used to shape and form metals often take a little more patience and practice but with care and practice outcomes that are accurate can be produced. Candidates' work shown on these pages uses different types of metal and the techniques used show the versatility of many metals.

- Take care at each stage of making and use cheaper modelling materials to develop your idea before using more expensive materials.
- The candidate in Sources **A** and **B** has used a card and wire modelling to produce an unusual candle holder (Source **C**) that allows ideas to be developed and modifications made where necessary.

A B C D

- Be careful that you avoid trying to increase the complexity of making at the expense of not fully completing your work accurately and to the highest standard.
- Sources **A** and **D** show candidates who have tackled a similar problem but used different materials to produce items.
- Source **D** has used a more difficult material to shape but has included a number of fabrication techniques – an applied finish is missing.
- Source **A** is neat and accurate but uses a single repeated fabrication process. The copper edges are smooth and it has been well polished.

Objectives

Develop and make a quality solution to a problem, choosing appropriate materials.

The examples on these pages are made predominently from metal.

links

Study sections 'Materials and components' and 'Processes and manufacture' to find more information about metals and the manner in which they can be formed, moulded and fabricated.

- Try to make your project as demanding as possible by perhaps using different metals and fabrication techniques. A complex and difficult making process allows access to higher marks.
- In both Sources **E** and **F** candidates have used the casting process but have produced items using different types of pattern.
- Source **E** has been produced to their pattern by using CAM (laser cutter) to produce the intricate shapes and numbers that are included in the flat backed pattern. Source **F** has used a more complex split pattern to produce a clamping device for a music stand.
- The candidate in Source **E** has produced an outcome using a manufacturing technique that requires a high degree of skill – making a pattern to be used in the casting process.
- The wooden and card pattern are tested to ensure that the pattern has the desired 'draft' to be removed from the casting sand.

E

F

- You must finish your final outcome to a high standard in terms of accuracy and the use of any necessary applied finish.
- The casting in Source **E** has been fettled and an applied finish (paint) has been successfully applied to certain sections. Parts of the casting have been machined and polished/buffed well.

links

Study the casting process in Casting on page 120. Pewter casting is a popular process that allows complex objects such as jewellery to be produced in school workshops.

Summary

It is necessary to produce a completed project that shows a high level of accuracy and a necessary applied finish.

Metals are available in different types, sizes and sections, and they offer a versatile range of manufacturing process.

15.3 Plastics

There are a number of common plastics available in schools that can be moulded, bent and formed. An increase in the availability of CAM has helped many candidates develop more intricate and complex designs that were very difficult to produce just using hand techniques. The different approaches used by the candidates have produced items that do not simply rely on CAM components but use a range of fabrication techniques that integrate other materials.

Objectives

Develop and make a quality solution to a problem, choosing appropriate materials.

The examples on these pages are made predominantly from plastics.

- Try to choose and use materials so that their material characteristics are utilised.
- The candidate in Source **A** has used a range of self-coloured plastic sheets to attract a target audience of young children. It has the advantage of being easily wiped clean.
- The candidate has produced a high quality item that might be made industrially using an injection moulding process.

A

B

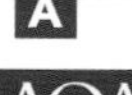

- Try to make your project as demanding as possible by perhaps taking an unusual approach to an undemanding problem.
- Source **B** shows a candidate who has taken a relatively simple project and used CAM to produce an intricate component difficult to manufacture by hand methods. Demand is increased by integrating this component with other materials to achieve a quality outcome.

- CAM can be used to help make batch components in a very short timescale.
- In Source **C** small chess pieces have been developed by cutting out and interlocking several pieces of plastic. This is an excellent way to produce components that may have been moulded in industry.

C

- Demonstrate as many skills as possible.
- The candidate in Source **D** has produced a quality outcome using a variety of manufacturing techniques.
- The final outcome is high quality as a result of 3D testing carried out in card (laser cut) and includes some high quality shaping and bending as well as a demanding aluminium casting.

D

Summary

The use of CAM with plastics offers an opportunity to rapidly manufacture items with excellent levels of accuracy and complexity.

A high level of accuracy is necessary and the quality of surface finish is especially important with plastics.

It can be difficult to manufacture using some plastic moulding processes in school workshops due to availability of complex and expensive equipment.

Quality checks carried out throughout the process can improve the accuracy of the final outcome and also help access improved grades.

15.4 Modern materials, electronics, mechanisms and smart materials

Smart or modern materials may allow you to produce a potentially different response in your design to the one you create when using more traditional materials. The completed projects used on these pages provide some examples of the different responses candidates have created in using electronics or mechanisms in their making. Mechanisms are often tried and tested several times in prototype form to ensure that the final mechanism included in a project works efficiently.

Objectives

Develop and make a quality solution to a problem, choosing appropriate materials.

These examples on these pages are made predominantly from smart or modern materials or include electronics or mechanisms.

- Electronic circuits can be used to increase complexity and demand in your project but only if they enhance the work. Their inclusion is not a substitute for poor standards in the manufacture of your project.
- The candidate in Sources **A** and **B** has used both electronics and CAM to produce a prototype MP3 docking station that shows evidence of a very high level of making.

links

Study Chapters 7 and 12 to find more information about electronics, mechanisms and modern/smart materials. Carefully consider ways that these may be incorporated into your design.

A

B

- Using an electronic kit, such as, radio, allows demanding projects to be tackled using a variety of materials and processes.
- Sources **C**, **D** and **E** show candidates who tackled a radio project and tried to produce an outcome with demand and complexity.
- Candidates who have used electronic kits in their projects have often successfully produced projects in the style of designers whom they have previously studied.

links

Study 3.1 Famous designers, to find more information about famous designers and the type and style of work they produced. Carefully consider ways that the information you have identified may influence your design.

C

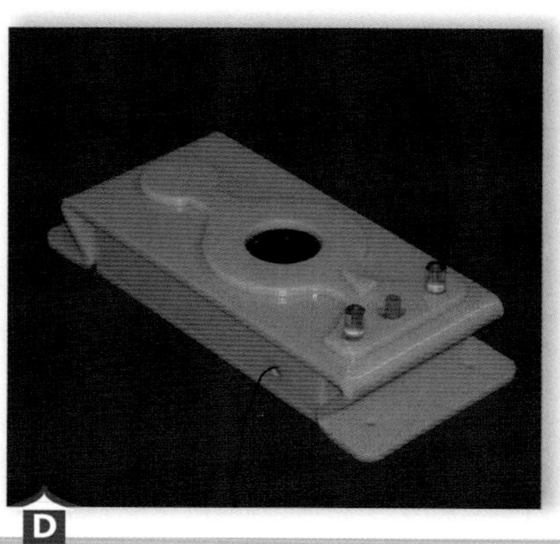

D

E

- Always make sure you finish your final outcome to the highest possible standard.
- The candidate in Source **F** has experimented with a simple mechanism that will be used in an automaton. Even though this is the prototype it is a quality outcome made with a high degree of skill to produce a working mechanism.

F

G

- At each stage of making try to ensure that any models and prototypes are made to the highest possible standards in terms of accuracy. Always include all your modelling/ prototypes in your design folio.
- In Source **F** each component was checked for accuracy before assembly and particular attention was paid to reducing friction in the mechanism.

- In both Sources **F** and **G** candidates have used similar mechanisms to produce items that convert rotary motion into linear movement.
- Accurately modelling different mechanisms is a very good way of making sure that an inclusion of a mechanism in your final project works efficiently.
- Source **G** was the first of several attempts to produce a 'snapping' action for the crocodile's mouth. Experimental work can often be helpful to develop an unusual mechanical movement.

links

Study Mechanical systems on page 132, and 12.2 Electrical systems on page 136 for further information regarding mechanisms.

Summary

Problems can be tackled in an innovative manner using a wide range of techniques to produce products that may include some mechanisms, electronics or smart materials.

A high level of accuracy is necessary in the production of products even when using mechanisms, electronics or modern materials.

16 Testing and evaluation

16.1 Testing and evaluation

The most important thing you need to know about testing and evaluation is that it is 'on-going' throughout your folder and not only something that you do right at the end.

Objectives

Understand a variety of methods of 'on-going' testing and evaluation.

Have a knowledge of a number of methods of final testing and evaluation.

'On-going' testing

There are a number of opportunities for testing and evaluating that can be done during the development stage.

Modelling

By producing a model you are able to quickly produce a 3D version of your design that can then be tested and evaluated by yourself and your target market.

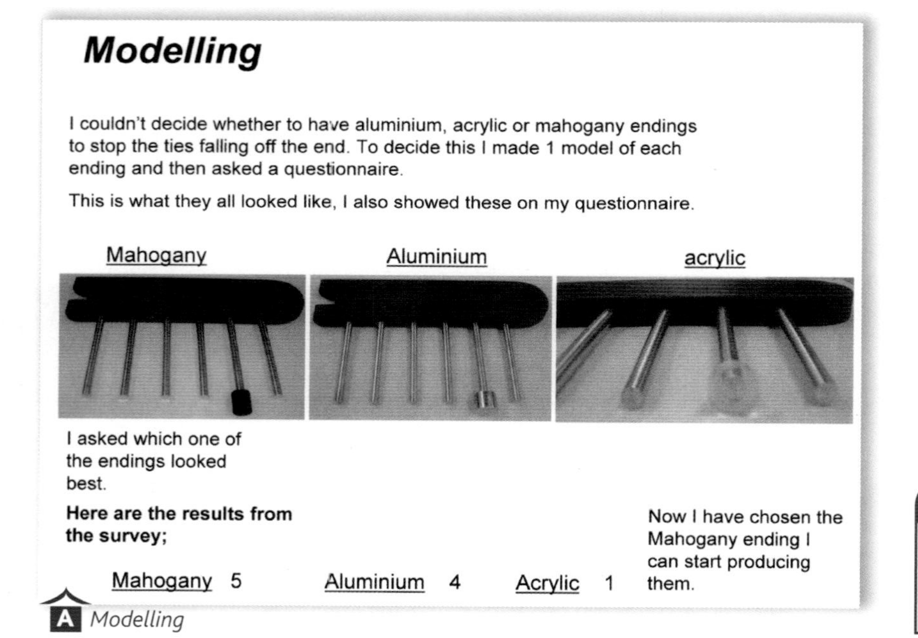

A *Modelling*

AC5 communication

Good use has been made of digital photography to show the student's models.

Survey/questionnaire

You can carry out a survey or produce a questionnaire to find out what your target market thinks of your modelled design.

Materials and finishes testing

You can look at a variety of materials and finishes; try them out to find out which material performs the best. Remember to record your results.

Manufacture testing

You can try out different methods of making your design to find out which method is the best. Again, remember to record your results.

'On-going' evaluation

Research

All aspects of your research should be evaluated and conclusions drawn up to influence your design specification.

Ideas

Each of your ideas should be evaluated to see how they perform against your specification.

Final evaluation

The final evaluation will allow you to evaluate the whole designing and making process.

Final evaluation v. specification

One of the best ways to start your final evaluation is by using your original specification. Take each point on your specification and ask it as a question of your finished project. Answer honestly; you get just as much credit for being critical of your finished project as you do for praising it.

AC5 communication

Good use has been made of Word to evaluate the final outcome against the original specification.

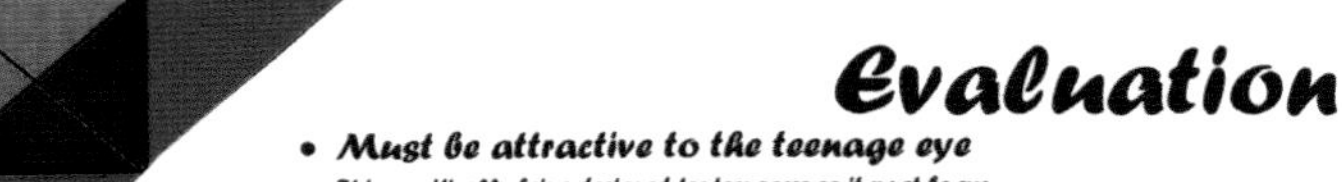

Evaluation

- **Must be attractive to the teenage eye**
  It is specifically being designed for teenagers so it must be eye-catching to this certain generation.
  Since I am a teenager myself I believe that my views are allowed to be taken into consideration and I think it is very attractive, but the fact that I have not done the football idea may have decreased its appeal to many male teenagers, but this means it will appeal to more females.

- **Must be a suitable size**
  It must be small enough to take around without it being a nuisance but must be big enough so it isn't hard to play.
  It's big enough to prevent it being difficult to play but with the aluminium feet it may not be as portable as it could have been.

- **It must be cost effective**
  It should be of good quality but it must be cheap enough to make, sell and get a profit out of it.
  Since I have used mahogany and mirrored acrylic it is not cheap but I think it has a very good quality-price ratio.

- **Must have no sharp edges**
  This is a safety precaution; it must not have sharp edges to minimize potential injuries.
  Since I have used Sanford edges there are no edges which could cause serious injury.

- **Must be secure**
  Even though it is portable, it still must be secure and stable. It needs to be secure enough to be able to play without the thought in the back of the mind that a slight knock can ruin the whole game.
  The magnets help to keep the top in place and the aluminium feet are perfectly sized so they do not wobble.

- **Must be ergonomically designed using anthropometric data**
  It must be designed so teenagers will be able to grasp the pieces with ease without knocking other pieces on the board.
  The space between the edge of a square on the board and the edge of a

- **Must be easy to use**
  It must be easy to get out of its compacted state, easy to play and then easy to put away. Also the quicker it can be taken out and put away the better.
  Unfortunately the feet cannot be folded away but the pieces are stored away in pockets which can be accessed very easily.

- **Must be well made/durable**
  It must of good enough quality to last quite a long time (about 7 years as this how long the teenage era lasts) without falling apart or breaking.
  Since all the mahogany has been coated with Danish Oil it is now waterproof which has greatly increased its durability and the acrylic board will last forever without eroding or losing shine.

- **Must be environmentally friendly**
  This includes making and selling. It must not be made using harmful materials/chemicals e.g. asbestos.
  Mahogany is a natural wood so it is no danger to the environment; acrylic does use crude oil to make but is not harmful once made; MDF can be carcinogenic but only in great amounts can it be of harm and aluminium is a naturally occurring metal.

- **Must be stylish**
  Teenagers will only buy this product if they think it is stylish. It could be themed or attractively designed.
  The mahogany edging gives it a quality look whilst the mirrored acrylic looks very impressive. Also the feet shine beautifully.

- **If it is being made out of wood, it must not splinter**
  If the product splinters it will not only not be very durable but it could also cause harm to people who into contact with it.
  The mahogany should not splinter easily as it has been finished with Danish Oil and MDF does not splinter at all.

B *Final evaluation v. specification*

Final survey/questionnaire

You can carry out a final survey/questionnaire to find out what your target market thinks of your finished project.

Final testing

You can now test your finished project to find out if it actually works. You could take a photograph of your finished project in use. If you are using e-Portfolio this is an excellent opportunity to use a video clip.

C *Final testing*

AC5 communication

Good use has been made of digital photography to show the testing of the final outcome.

Modifications during manufacture

There are a number of reasons why you may wish to modify your design during the making stage.

You may well have thought of a better design for a part of your project while you are making it. Go ahead, make the changes, but make sure that you record the changes and why you made them. A photograph at this stage will help.

You may well have been forced to change your design due to difficulties sourcing materials, material performance or manufacturing difficulties. This is not a problem. Again, go ahead, make the changes, but make sure that you record the changes and why you made them.

If you are truly challenging yourself then you will, at some stage, have a problem when you are making your project. There is an old saying that goes, 'The man who never made a mistake never made anything.' It's true! Tell us what went wrong and, more importantly, tell us what you did about it.

... and finally

If you were to begin this project all over again, what would your 'Mark 2' look like?

If you are a true designer you will have been thinking about this throughout the project. Produce a sketch of your 'Mark 2', clearly explaining the changes and why you would make them.

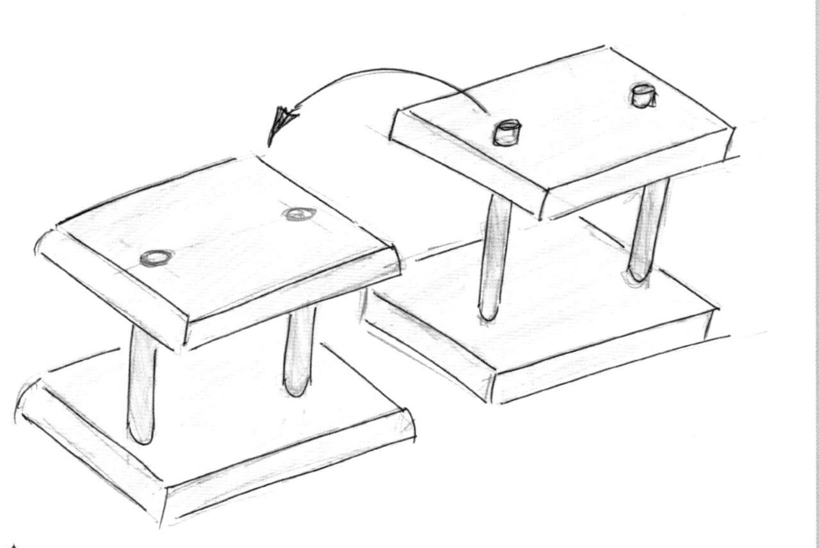

D *Mark 2*

AC5 communication

Good use has been made of 3D sketching to clearly show further improvements.

Summary

There are a number of ways of carrying out 'on-going' testing. This could include: modelling, survey/questionnaire, materials and finishes testing, manufacturing testing.
All aspects of your research should be evaluated.
All ideas should be evaluated against the specification.
Your finished project should be evaluated against your specification and fully tested.
Produce a report on any modifications that you made during manufacture.
Explain any changes you would want to make if you were given the opportunity to do your project all over again.

17 Frequently Asked Questions

Controlled Assessment

Can I make anything I want for my controlled assessment?

You must talk to your teacher about your ideas for your controlled assessment. It must be one that is on the list of approved controlled assessments given to your teacher by the AQA. You and your teacher must decide if the controlled assessment you choose is the right one for you. You should be able to complete it but it should also test your abilities to the full.

How long have I got to design and make my controlled assessment?

You have approximately 45 hours in which to produce your controlled assessment. This includes any time you spend working on your project in after school clubs etc.

Can I work on my controlled assessment at home or on my work experience?

Your controlled assessment must be done under informal supervision. This means that most of it will be done in school with your teacher present.

How big should my design folder be?

AQA recommend approximately 20 pages of A3 paper (or the A4 or electronic equivalent).

I am very good at making in wood; can I produce my project in just one material?

The emphasis is on showing your practical skills. The choice of material or materials is entirely up to you and your teacher. However, small projects that show lots of high levels of skill, both in making and finishing will score much higher than large projects that only show a limited range of skill.

We have a lot of CAD/CAM equipment at school. Can I completely design and make my controlled assessment by computer?

The use of CAD/CAM is encouraged by AQA. However you must challenge yourself to the same level as a student who uses purely traditional methods.

What do I do with the preparation sheet?

The preparation sheet tells you what the theme of the design question will be. It is given to you so you can do some research, on the given topic, before the examination. Make sure that you have looked at existing design, the materials from which they are made, the finish that has been applied to them and how they have been manufactured.

Glossary

6 Rs: six words beginning with the letter R. Each word describes action that can be taken to reduce the environmental impact of products.

A

Active disassembly: taking apart a product semi-automatically, by using smart materials for the fastenings. The fastenings release each part of the product at a pre-set temperature.

Adhesives: types of glue.

Aesthetics: the features of a shape that make it look good.

Allen key: an L-shaped tool for undoing screws with a hexagon-shaped socket in the head.

Alloys: metals formed by mixing together two or more metals to produce a new metal that has improved characteristics/ properties.

Annealing: the softening of metal by heating to a specific temperature and then allowing to cool.

Anthropometrics: measurements of the human body.

Assemble (in CAD): place together separate parts of an object drawn in computer-aided design.

B

Biodegradable: a material that breaks down naturally with time. Sunlight, rain or bacteria could break down the material.

Biodegrade: break down naturally, decompose, rot (for example, timber).

Buffed: mechanically or hand-polished to produce a high-quality, shiny surface.

C

CAD: computer-aided design.

CAM: computer-aided manufacture-machines that are computer numerically controlled (CNC).

CE mark: a declaration by the manufacturer that their product meets the requirements of the applicable European Directives.

Client/user/consumer profile: a description of the typical person or people who will use the product.

COSHH: Control of Substances Hazardous to Health.

Constraints: things that impose limits and boundaries to aspects of a product such as its size (size constraints) or the number of features (cost constraints).

Countersunk: a hole that is recessed, to allow the head of a countersunk screw to be driven level with the wood surface.

Culture: the beliefs, fashion, music, likes and dislikes of certain groups within society.

D

Datum edge/line: this is used to make all measurements from. It ensures accuracy and prevents errors accumulating.

Deburring: removing the sharp edge from a piece of metal by drawing a file backwards and forwards.

Develop (ideas): simplify, refine, clarify, modify and decide constructional methods to improve your design.

Disposable: a product that is designed to be used a limited number of times before being thrown away.

Draw filing: smoothing the edge of metal or plastic by drawing a file backwards and forwards.

E

Electrical circuit: a number of electrical components connected together to form a functioning electrical product.

Electrical component: an electrical part that forms an electrical circuit, for example, batteries, switches and LEDs.

Electrolysis: the process of coating a metal by placing it into a solution of electrolyte and passing an electric current from the donor metal to the parent metal.

Ergonomic: something that has been designed to allow people to work efficiently by making it comfortable and user-friendly.

Ergonomics: the study of people in relation to their working environment, in order to maximise efficiency.

Evaluate: make judgements about how successful something is.

Exclusive design: designs that are accessible only by particular members of society.

Exploded view: a 3D sketch or drawing where the parts of the product have been pulled apart and drawn in line with the parts to which they need to fit.

F

Fabrication: the joining together of pieces, whether or not they are the same material.

Ferrous: group of metals that contain iron and varying amounts of carbon. They are normally magnetic.

Filler rod: an alloy that is used to bond two metals together as they are heated.

Flowchart: a list of processes organised into the best order. Decisions need to be made at stages in the flowchart. Arrows show the route through.

Flux: a chemical cleaner used to clean metal as heat is applied.

Forging: a traditional process that uses a hammering action provided either by hand or from a machine to create a variety of shapes.

Form: form deals with the shape of a product.

Function: function deals with how a product works.

G

Galvanised: a coating of zinc applied to steel to stop it from rusting.

H

Hardwood: timber that tends to be from slow growing, broad-leafed trees.

I

Inclusive design: designs that are accessible by all members of society.

J

Jig: an aid to fast, accurate and repeatable manufacturing operation.

K

Keyed: where a surface is roughened to improve the strength of a joint when two surfaces are stuck together.

Kit car: a self-assembly car, put together by amateur car builders.

Kitemark: the mark which shows that a product has been tested to meet international standards

Knock-down (KD) fitting: a component that allows rapid assembly and disassembly, without damage to the parts being joined or separated. The fitting often has two parts, with one screwing or locking into the other.

L

Laminating: the process of bonding two or more layers of material together to form a thicker and stronger section.

Landfill: a large hole in the ground that is filled with rubbish that is not being recycled. Sometimes old quarries are used for this.

M

Machine tools: larger machines in a school workshop that are normally fixed to either a bench or the floor, are more powerful, and should have a range of safety devices to protect the user.

Maintenance: cleaning, adjustment, lubrication or replacement of parts of a product, to allow it to continue to function correctly.

Market research: collection of information from likely users of a product, to find their needs and what they would like

Mechanical advantage: the way in which a machine makes things physically easier to do.

Mechanical component: a mechanical part of a larger system or product.

Mechanical properties: properties of materials including strength, hardness, density, durability, toughness/brittleness, malleability, ductility and elasticity.

Mechanical system: an assembly of mechanical components that form a machine.

Microelectronics: the miniaturisation of electronics making products smaller and smarter

Modernist: ideas from a period of design history around the turn of the 20th century. Modernism replaced traditional, extravagant products with very usable, stylish, original and exciting designs.

Molten: the state when a metal has turned to liquid by heating.

Moral issues: the right or wrong of an action.

Mould: a 3D template used to create a required shape. It must be made from a material that can withstand both heat and slight pressure.

N

Nanotechnology: the technology used to rearrange individual atoms to create new, improved materials, systems and devices.

Non-ferrous: group of metals that do not contain iron.

O

Obsolescence: lack of appeal to consumers because something goes out of date and better products become available.

Obsolete: out of date, no longer required.

Ore: a solid, natural material from which metal can be extracted.

P

Patent: these protect the features and processes that make things work. This lets inventors profit from their inventions.

Patina: a thin coloured film that forms naturally on the surface of some metals.

Pattern: an exact copy of the shape required. Certain shapes are made slightly larger than the required finished size to allow for shrinkage as the material solidifies.

Pendant: a piece of jewellery, usually worn hanging from a chain around the neck.

Personal protective equipment (PPE): for example goggles to protect the eyes.

Physical properties: properties of materials including fusibility, conductivity and environmental friendliness.

Polymorph: a smart material that is easily formed when heated, and solidifies when cool.

Prototype: a model of a product that is used to test a design before it goes into production.

Q

Quality assurance: a complete system of quality control checks and procedures throughout the manufacture of a product.

Quality control: a check made to ensure that a component meets the specification, for example, correct size, shape, colour.

R

Rendering: adding colour to a sketch to help show the materials and textures of surfaces.

S

Softwood: timber from quick growing conifers.

Sustainable: something that can be replaced or reused/recycled indefinitely.

Sustainability: the ability to keep making or using a product without excessive damage to the environment

Sweatshop: a factory where people work for long hours, in poor conditions and for little pay.

T

Target market/target group: the people who, it is hoped, will buy and use a new product.

Template: a device to mark out components quickly and accurately.

Thermochromic: having the ability to change colour as the temperature is varied.

Thermoplastics: become soft and pliable when heated and can be reheated as often as required. As they cool they set again.

Thermosetting plastics: soft and pliable the first time they are heated but a chemical change takes place on cooling and they then become rigid, non-flexible and cannot be reheated and changed.

Tolerance: the amount of error that can be allowed and is often given as ± mm. It is more important in some products than others.

U

Unique: something that is a one-off; there is nothing else the same as it.

Veneer: a thin section of timber that is cut from a log and then used to produce plywood, or is glued on top of a cheaper material.

W

Wasting: the mechanical removal of unwanted material by use of tools or machinery that use a cutting action.

Wood joint: a system used to join two or more pieces of wood together that might have some mechanical strength and is likely to be permanently joined using an adhesive.

Wood turning: a manufacturing method that uses specialist tools to shape wood that is being spun or rotated.

Work hardening: as a result of deformation of a metal there is an increase in hardness that may eventually cause fractures.

Index

Key terms, and page(s) where defined, appear in **bold type**.